| 中英双语版 |

Tender Thistles

温暖的荆棘

毕淑敏写给年轻人的心理学故事

毕淑敏–作品
朱虹–译

天地出版社 | TIANDI PRESS

图书在版编目（CIP）数据

温暖的荆棘：汉英对照 / 毕淑敏著；朱虹译. —
成都：天地出版社，2017.10（2018年重印）
ISBN 978-7-5455-3063-6

I. ①温… II. ①毕… ②朱… Ⅲ. ①青年心理学—
通俗读物—汉、英 Ⅳ. ①B844.2-49

中国版本图书馆CIP数据核字（2017）第209407号

温暖的荆棘

出品人　杨　政
著　者　毕淑敏
译　者　朱　虹
责任编辑　张秋红　沈海霞
封面设计　思想工社
电脑制作　思想工社
责任印制　葛红梅

出版发行　天地出版社
（成都市槐树街2号　邮政编码：610014）
网　址　http://www.tiandiph.com
http://www.天地出版社.com
电子邮箱　tiandicbs@vip.163.com
经　销　新华文轩出版传媒股份有限公司

印　刷　河北鹏润印刷有限公司
版　次　2018年1月第1版
印　次　2018年7月第3次印刷
成品尺寸　155mm×220mm　1/16
印　张　18
字　数　286千字
定　价　39.80元
书　号　ISBN 978-7-5455-3063-6

咨询电话：（028）87734639（总编室）
购书热线：（010）67693207（市场部）

你不能要求拥有一个没有风暴的人生海洋，

因为痛苦和磨难是人生的一部分。

一个没有风暴的海洋，

那不是海，是泥塘。

序言

有许多作家有病

王蒙

有许多作家有病，例如陀思妥耶夫斯基的癫痫，契诃夫的肺结核，杜勃洛留波夫的瘰疬，李贺与子规（日本俳句诗人）的咯血与夭亡，杰克·伦敦、海明威、川端康成、三岛由纪夫的自杀等。

Preface

Writers are Sickly People

Wang Meng

All the way down history, many writers have been sickly people.Take Dostoevsky, for instance. He had epilepsy. Chekhov had tuberculosis. Dobroliubov had boils tuberculosis of lymph nodes. Both the Tang poet Li He and the Japanese writer Masaoka Shiki had hemoptysis and died of tuberculosis at an early age. As for Jack London, Ernest Hemingway, Kawabata Yasunari, Yukio Mishima and others, they had all died by their own hands.

还有一些作家的疾病没有这样众所周知，他们的疾患更多在精神方面，尽管他们可能是很有成就的作家，但是一读他们的作品，那种偏执，那种自恋自吹、自我膨胀，那种迫害狂，那种抑郁加躁狂的无可控制，那种神经兮兮和白昼见鬼，都令人嗟叹。这里边确实有天才，生老病死的苦痛需要通过这些作家的手生发成花花果果，不能因为疾患就忽视他们的天才，但也不能因天才就不敢正视他们的疾患。

我也希望作家中有真正的医生，而不仅是有过成名前行医的记录。疾患越多，对医生的期待就越大。

咱们这里有一个毕淑敏。解放军的卫生员、医科大学生、内科主任、小说与散文家、心理学硕士、心理咨询专家，从小的好学生、好孩子，不那么另类，而是符合主流价值的有为者。

Then there were writers whose illness had not gotten around through the grapevine. Their illness had been mostly on the psychological side. There is no denying their accomplishments as writers, but the minute you pick up one of their works, you cannot help being staggered by their uncontrollable fits of:

pig-headedness, self-conceit, braggadachio, self-aggrandizement, persecution-mania, depression, uncontrollable violence, and more.

There is no denying their genius.

But they see ghosts in broad daylight, and are completely off their rocker.

The pains of living and dying are turned into colorful art through their hands.

But we must not turn a blind eye to their medical problems because of their genius.

Other writers had health problems too, not known to the outside world, mostly mental problems. But again, there is no denying their genius.

They are indisputably great writers, but when you read them, you cannot help being struck by their perversion, self-conceit and persecution mania.

They are indeed geniuses: the pains of living and dying dazzle through their writings.

We should not discard them as writers on the grounds that they themselves were sick.

On the other hand, we should not turn a blind eye to their sickness on the grounds of their literary genius.

I wish there were practicing doctors within the ranks of practicing writers—not writers who had dabbled in medicine before they became famous as writers. Disease being rampant, people's expectations from doctors are rising daily.

We do have a writer among us—Bi Shumin. She had been a PLA clinical worker, a medical student, head of the division for internal medicine, novelist and essayist, holder of an MA degree in psychology, and a counseling psychologist. A model student and good daughter since childhood—never a freak—in a word, she is a person with outstanding accomplishments keeping within mainstream values.

医学是科学。医术是高科技。医心是天使的心、菩萨的心，济世救危，助弱扶伤，杨枝净水，慈悲为怀。爱心是主干，责任是永久，使命是奉献。

小说与散文是人的故事和情感，体贴入微，心界万象，人生经验，苦辣酸甜，再加生花的妙笔，成精成色的语言。

心理咨询是新兴产业，是人的悲哀与软弱的证明。生活越是复杂，社会越是精密，节奏越是紧急，竞争越是激烈，心理的负担就越多，就越需要医生的专业，更需要医生的抚慰、疏浚与理解。

Medicine is a branch of science.

Practising medicine is high technology.

Doctors are angels with hearts of Buddha,

Supporting the weak and ministering to the sick.

Following their hearts, they are full of compassion;

Aware of their responsibility, their mission is to give.

Novels are narratives of human feelings told in vivid detail,

The experience of a lifetime, the grip of emotion sweet and bitter,

Passing through the pen of scribblers, they flower out in magic writing.

Psychological counseling is a new line of business,

A proof of the human sadness and frailty.

As life gets increasingly complicated,

As the society is increasingly knit,

As the pace of life is urgent,

As competition is increasingly tense,

So is our mental burden increasingly heavy,

And all the more do we need the doctor's concern, dispensation, and understanding.

读了毕淑敏著的文章，你觉得这些都有了。它减轻你的痛苦，虽然时下兴的是绞紧你的肉和心。它开阔你的心胸，虽然时下兴的是鼻子底下与脐下三寸那点事死缠不休。它写得细腻而又晓畅，虽然时下兴的是故作艰深。它拳拳而又眷眷，虽然时下兴的是恶毒与残酷的刺激。它合情合理，虽然时下兴的是大言爆破。它告诉你一加一还是等于二，虽然时下兴的是一加零等于三万八千八。它娓娓动听，虽然时下兴的是对于阅读的颠覆。它竭力教给你活得好一点，快乐一点，善良一点，健康一点，光明一点，虽然高烧、病态、梦呓也是一种吸引眼球的行销策略。

如果我有亲戚朋友，我有孩子，我有孩子的朋友和老师，如果他们都是普通的与正常的人，我都会推荐他或她读这本书。至少，在读一些令人犯病的书之外，也不妨，也应该，也亟须读一点儿教人健康的文学书。

Read Bi Shumin's stories,

And you will find it all there.

The stories relieve your suffering,

Though nowadays it is cool to put you through the grinder.

Her stories put your heart at ease,

Though nowadays the focus is on that boring thingamy between your legs.

Smoothly her story flows in colorful detail,

Though nowadays pseudo-profundity is all the rage.

She is tough with a gentle touch,

Though nowadays cruelty and malice has the upper hand.

She is cool and reasonable,

Though nowadays raging and ranting is what you hear.

She assures you that one and one makes two,

Though nowadays the total runs into hundreds of thousands.

Her writing is smooth and elegant,

Though nowadays to read is to subvert.

She urges you—enjoy life; be happy; be kind; be healthy; be sunny,

Though nowadays high fever, paranoia and nightmares are good strategies to catch your eye.

If I have friends and relatives, if I have children and they have friends and if they were ordinary normal people, I would recommend them to read this book.

As an alternative to reading books that make you sick, or as a harmless concession, or rather as an urgent need, one should read works of literature that make you happy.

自序

书里藏着两双手

毕淑敏

本书英文译者朱虹老师的手，温热而柔软。真是难以相信，这样一双看似普通的手，写过大量极有创见的英美文学评论。尤其是用准确简练又不乏幽默的笔触，将大量中国文学翻译成英文，并得到英语世界主流文学界的赞赏与认同，她的翻译还不止一次成为博士论文的课题。

朱虹老师由衷热爱翻译事业，一直兢兢业业，她以英语的思维，追求“神似而非形似”的翻译风格，不是亦步亦趋地简单直译，而是在力求准确的同时，不失灵性和变通，严谨中饱含风趣，让读者领会更加原汁原味的英文，体会纯正的英语之美。这样一位声震海内外的翻译大家，怎么会在 84 周岁的耄耋之年，翻译我的这本小书呢？

她的初心，是为了留给自己的孙女们一件精致隽远的礼物，便倾注了满腔心血来做这件事。译文反复推敲修改，精益求精，达到感人至深的地步。

她老人家原想只把此书作为自家的珍宝，亲人相传。不想此译稿被天地出版社得到，将它出版，于是更多的读者也有了珍藏它的机缘。

Foreword

Two Pairs of Hands Hidden in a Book

Bi Shumin

The hands of Zhu Hong, the translator of this volume of stories, are soft and warm.

It is hard to imagine that this pair of hands had written articles on English and American literature with sharp critical acumen.

More amazing is the fact that Zhu Hong had translated outstanding works of Chinese literature into English. Her translations, precise, succinct, sometimes with a dash of humor thrown in, have been approved of and accepted in the English-speaking world of academia, while at home they have more than once served as topics for Ph.D theses.

Zhu Hong works hard and is deeply devoted to her chosen field. True to her ideal of translation not as a mere formation of words but as a means of imparting the spirit of the original, Zhu Hong's works of translation have managed to offer her readers a taste of the English language in all its beauty and flexibility.

Now why did this famed master of translation, at the venerable age of eighty-four, bother to translate these short stories of mine? It turned out that Zhu Hong had planned to translate this collection of my stories as a gift to all her granddaughters, a gift to be handed down as an heirloom within the family. Thus she threw herself into the job of translation, sweating over every word in pursuit of perfection. However, the project was discovered by Tiandi Press, and thus destined to be released for a larger readership.

文字之中，也藏着我的一双手。我这双手，没什么特殊，只是指掌比较大。当年我做外科医生的时候，如果第二天有我出台做手术，护士长会特别叮嘱准备手术器械的新护士说，一般女医生的手套6号半到7号即可，毕医生要用7号半的大手套。

现在上了点年纪，因长年写作，多个手指发炎变粗，估计再要去做手术，没准要用8号的手套了。

说完了手，说说我的心意吧。我向你保证，书中文字，都是我秉着当年执手术时的责任感，以治病救人为无尚光荣的医生之德郑重写下，力求它们传达暖意和富有力量。

感谢天地出版社编辑们的辛勤劳动。

亲爱的读者，这样流畅美好伏贴而生动幽默的英文译本，真是难得的宝贝。赶紧把它收藏起来吧。

我和朱虹老师的手，藏在书中，等着和你打招呼呢！

My own two hands are also hidden in the text of this book.

There is nothing special about my hands, except that it is larger than the usual for women of my age. Back in those days when I was a surgeon and whenever there was a surgery planned for the next day, the nurse in charge of surgical instruments would always order gloves of size 7.5, explaining to her staff that size 6—7 was the usual for women doctors, but that Doctor Bi Shumin needed 7.5!

By now, I have added some years to my age, not to mention having spent those years writing nonstop, and most of my fingers are swollen with inflammation. If I am to perform surgery now, I guess I will need size 8 gloves.

Now that we have dealt with hands, let us move on to the heart.

Let me assure you, readers: every word here in the book is sprung from my sense of responsibility as a practicing surgeon. All of them were written down conscientiously, word by word, out of a sense of pride in being a doctor dedicated to saving lives, and were all written in the hope that those words, imbued with strength, would send out warmth!

Many thanks to the editorial staff of Tiandi Press for their hard work!

Dear readers, with such smooth flowing English, faithful to the original, bubbling with humor, such a book is a rare acquisition. Snatch it!

My hands and Zhu Hong's hands—are hidden inside the book, waiting to greet you!

C O N T E N T S

目　录

C O N T E N T S

目 录

C O N T E N T S
目 录

自助看医生

儿子长得比我高了。一天，我看他有点儿打蔫儿，就习惯性地摸摸他的头，在这一瞬间的触摸中，我知道他在发烧。

“你病了。”我说。

“噢，可能是病了。我还以为是睡觉少了呢。妈妈，我该吃点儿什么药？”他问。

我当过许多年医生，孩子有病，一般都是我在家里给治了，他几乎没有去过医院。这次，当我又准备在家里的储药柜里找药时，却突然怔住了。

“你长大了，你得学会看病。”我说。

“看病还用学吗？您给看看不就行了吗？”他大吃一惊。

“假如我不在家呢？”

“那我就打电话找你。”

“假如……你找不到我呢？”

“那我就……找我爸。”

这样逼问一个生病的孩子也许是一种残忍。但我知道，总有一天他必须独立面对疾病。既然我是母亲，就应该及早教会他看病。

“假如你最终也找不到你爸呢？”

“那我就忍着。反正你们早晚会回家的。”儿子说。

“有些病是不能忍的，早治一分钟是一分钟。得了病最应该做的事是上医院。”

A do-it-yourself Visit to the Doctor

By now my son is taller than I am. One day I saw that he was mooning about, listless. I ran my hand over his tousled hair. Goodness, his head was burning hot. He was running a fever.

"You are sick," I told him.

"I suppose so," he mumbled, "I thought that I was just missing sleep. What pills shall I take, Mum?"

I had been a practicing doctor for many years and had always treated him myself. He rarely ever went to hospital.

I rummaged through the medicine closet. But suddenly a thought struck me: "You are a big boy now. You must learn how to make a visit to the doctor."

"Learn how to see a doctor?! Why should I do that? Can't you treat me?" He acted shocked.

"Supposing I am not around?"

"I'll ring you up."

"Supposing I am not available?"

"I'll call Dad."

It seems mean to bother a sick child like that, but sooner or later, he must deal with his own illnesses. Being his mother, it was high time I taught him the basics.

"Supposing you can't find your dad?"

"I'll grin and bear it. You guys will be home sooner or later."

"Some problems are urgent. Every minute counts. When sick, one goes to see the doctor."

“妈妈，您的意思是让我独自去医院看病？”他说。

“正是。”我咬着牙说，生怕自己会改变主意。

“那好吧……”他摸着脑门，不知道是虚弱还是思考。

“你到街上去打车，然后到医院，先挂号，记住，要买一个病历本。然后到内科，先到分诊台，护士让你到几号诊室你就到几号，坐在门口等。查体温的时候不要把人家的体温表打碎……”我喋喋不休地指教着。

“妈妈，您不要说了。”儿子沙哑着嗓子说。

我的心立刻软了。是啊，孩子毕竟是孩子，而且是病中的孩子。我拉起他滚烫的手，说：“妈妈这就领你上医院。”

他挣开我的手，说：“我不是那个意思。我是说我要去找一支笔，把您说的看病的过程记下来，我好照着办。”

儿子摇摇晃晃地走了。从他出门的那一分钟起，我就开始后悔了。我想我一定是世上最狠心的母亲，在孩子有病的时候，不但不帮助他，还给他雪上加霜。我就是想锻炼他，也该领着他一道去，一路上指点指点，让他先有个印象，以后再按图索骥。这样虽说可能留不下记忆的痕迹，但来日方长，又何必在意这病中的分分秒秒呢？

“Are you saying that I should go to the hospital on my own?”

“That’s exactly what I mean!” I gritted my teeth, determined to be firm.

“Okay,” he said, as he ran his hand through his hair, either feeling ill, or just thinking.

“You take a taxi and get to hospital. Register, pay for a booklet for your medical record, head for the department of Internal Medicine, make sure you get to the right sub-division, wait to be called by your registration number, stay near the door to make sure you don’t miss your turn, don’t drop the thermometer when checking your temperature....”

“That’s enough, mom,” he announced in a hoarse voice.

My heart contracted in pain. He was but a child, after all, and a sick child to boot. I grasped his feverish hand and said: “Come on, Mum is taking you there!” He extracted his hand from my grasp and said: “That’s not what I mean, Mum. I am looking for a pen to write down the steps to go through for seeing a doctor, and I’ll follow through.”

He walked away unsteadily. The minute he was out the door, I was sorry. I felt myself to be the worst mother in the world, refusing to help her sick child and making it worse for him.

My idea had been to let him do things on his own, but I should have been by his side at the start. I should have shown him the way and given him time to remember the steps to be taken and get a sense of the whole process. To be sure, taking him there will not leave as deep an impression on his memory as letting him strike out on his own. But there is all the time in the world, why rush things while he is still sick!

时间艰涩地流动着，像沙漏坠入我忐忑不安的心房。两个小时过去了，儿子还没有回来。虽然我知道看病是件费时间的事，但我的心还是疼痛地收缩成一团。

虽然我毫无疑义地判定儿子患的只是普通感冒，如果寻找适宜锻炼看病的病种，这是最好的选择，但我还是深深地谴责自己。假如事情重来一遍，我再也不让他独自去看病了。这一刻，我只要他在我身边！

终于，走廊上响起了熟悉的脚步声，只是较平日拖沓。我开了门，倚在门上。

“我已经学会了看病。打了退烧针，现在我已经好多了。这真是件挺麻烦的事。不过，也没什么大不了的。”儿子骄傲地宣布，然后又补充说：“您让我记的那张纸，有的地方顺序不对。”

我看着他，勇气又渐渐回到心里。我知道应该不断地磨炼他，在这个过程中，也磨炼自己。

孩子，不要埋怨我在你生病时的冷漠。总有一天，你要离我远去，独自面对生活。我预先能帮助你的，就是向你口授一张路线图，它也许不那么准确，但聊胜于无。

Time passed by so quickly, like an hourglass ticking away in my heart. Two hours passed by, and he was still not back. I knew that a visit to the doctor was time-consuming, but my heart still contracted painfully.

It was obviously a common cold, a perfect case for a child to practice going to hospital on his own. But I still blamed myself. If Time could wind back, I would never allow him to go to hospital alone. At that moment, I wanted nothing more than to have him beside me.

Soon, familiar footsteps sounded in the corridor, though not as crisp as they used to be. I opened the door, and there he was, leaning against the door frame.

"I know how to see a doctor. I got a shot and my fever is down. It was a lot of fuss but no big deal," my son announced proudly.

"By the way," he added, "some of the items on your list are out of sequence."

I looked at him, and gradually regained my courage. I know that he will have to go through a lot more tempering. Me too.

Don't blame me for deserting you when you had a cold. You will leave home sooner or later, son, and start the journey of life on your own. All I can give you is an oral map. It might not be very precise, but it is better than nothing.

谁是
你的闺蜜

某天，我看到工作人员正在清理一堆小山似的硬币，好像是哪个孩子当场砸碎了他的宝贝扑满。我很奇怪，心理机构不是超市银行，似乎不应该收集如此多的硬币。助手们都很尽职，平常绝不会在业务场所处理私事，看来这些硬币和工作有关。我实在想不明白：硬币和心理咨询有何关系？

助手看我纳闷，就说："这是一个孩子交来的预约咨询费用。"我一时愣怔，心想，孩子的钱，是不是应该减免？助手看我不说话，以为我是在斟酌钱的数量，就说："这是那个孩子所有的钱，我打算自己帮她补足。"

我说："钱的事，咱们再说。我想知道孩子是跟着谁来的。"

按照惯例，孩子的问题，都是父母发现后焦虑不安地领来求助。

助手说："这孩子是自己来的，用压岁钱来付费，父母根本不知道她要来看心理医生。"助手说着，把她的登记表递过来。

工工整整的字迹填写着：张小锦，女，十三岁，本市××中学初一年级学生……

见到张小锦的时候，我吃了一惊。本以为这么敢作敢为挺有主意的孩子一定人高马大，却不料她十分瘦小，穿着橙色校服蜷在沙发中，好像一粒小小的黄米。

我说："你遇到了什么事情，需要我们的帮助？"

Best Buddies

One of the staff was sorting out a pile of coins scattered on the floor, as if a child's moneybox had been overturned. I was surprised. The Center for Psychological Counseling, being neither a bank nor a supermarket, has no reason to be deluged in coins. I could not make out how or why these coins got there. My assistant explained that a child had paid in coins for counseling. Should we reduce charges for a child? My assistant, seeing my dilemma, said: "Here is all she had in the moneybox, I was going to make up the difference." I said the money can wait, but I need to know who came with her. Usually it is the worried parents lugging over their problem child. My assistant said that the child came alone and had paid the fee with her New Year gift money. Her parents are in the dark, she added as she handed over the form that the child had filled. A row of neatly written characters revealed that the visitor was Zhang Xiaojin, female, thirteen years old, first year junior high at a local school in town.

I was shocked when I saw Zhang Xiaojin in person. Since she had taken things into her own hands, I had assumed that she must be big and tall. What I saw was a wisp of a girl in yellow school uniform, curled up in the depth of the sofa, like a little beanie.

I asked: "What is your problem, how can I help?"

瘦小的张小锦说起话来嗓门挺大，音调喑哑，有点像张柏芝，仿佛轻巧的身躯里藏着一根摔裂的长笛。张小锦咬牙切齿道：“我请你帮助我——除掉我妈的朋友！”

我着实被吓了一跳。这个开头，有点像黑帮买凶杀人。我说：“你很恨你妈妈的朋友？”

张小锦说：“那当然！请你千万不要将我的话告诉任何人。你要发誓，永远不能说。”

这可让我大大地为难了。就算她是一个孩子，如果她图谋杀人，我也要向有关机构报告。如果我拒绝了张小锦的要求，她很可能就拒绝和我说知心话了，帮助便无从谈起。我避开话锋，慢吞吞地回答：“你能告诉我，你说的‘除掉妈妈的朋友’是什么意思吗？”

“除掉”通常是血腥的。警匪影片中将要杀死某个人的时候，匪徒们会窃窃私语，吐出这个词。张小锦回答说：“我的‘除掉’就是让这个朋友离开我家！不要和我妈没完没了说个不停，让我妈多拿出一点时间来陪我，遇事别老听这个朋友的，也和我聊聊天，也听听我的想法……”

原来是这样！在张小锦的词典里，“除掉”并不是杀死，只是离开。我稍稍松了一口气，说：“张小锦，看来你妈妈和你交流不够，你对此很有意见啊。”

张小锦遇到了知音，直起身板说：“对啊！我妈有什么心事，只和朋友说，不和我说。我们家的事，是和她朋友关系密切啊，还是和我密切啊？”

张小锦黑亮的眼珠凝神盯着我，目光中带出急切和哀伤。

我立即表态：“你们家的事，当然是和你关系最密切了。”

Zhang Xiaojin's voice was loud and husky, as if issuing from a cracked bell hidden within that fragile little body. She gnashed her teeth and stated: "I want you to get rid of my mother's buddy!"

I was shocked. She sounded like a gangster, hiring someone to deal with a nuisance.

"You hate your mother's friend?"

"Of course I do! But no one must know! You must swear never to betray me!"

I was caught in a dilemma. If I have information about someone—even a child—contemplating murder, I am bound to report to the authorities. On the other hand, if I refuse Zhang Xiaojin's terms, she might not confide in me, and then it would be impossible for me to help her. I got around the dilemma by mumbling: "What do you mean by *getting rid of* your mother's friend?" "Getting rid of" is gangster language uttered in undertones, often heard in movies when referring to murder. Zhang Xiaojin answered my question by saying, "What I mean by *getting rid of* is for this 'friend' to leave my family alone! To stop chattering with my mom! To leave my mom alone and let my mom spend more time with *me*! To let my mom talk to *me*, and listen to what *I* have to say!"

So that's it! In Zhang Xiaojin's dictionary, "getting rid of" does not mean murder, but just to get somebody out of the way. I heaved a sigh of relief, saying "Xiaojin, you feel that you and your mom are not close enough, and you are not happy about it?"

Finding a confidante in me, Xiaojin sat up and asserted: "Yes, that's it! When my mom has a problem, she talks to *her*, not to me! Now tell me, should our family affairs be closer to a friend, or closer to *me*?" Zhang Xiaojin held me transfixed by her gaze, urgent and appealing.

I immediately responded, "Certainly your family affairs affect you more closely!"

这让张小锦很受用，她说："对啊！那个朋友一天到晚老缠着我妈，让我妈离婚，破坏我们家的和睦！"说着，她长长的睫毛湿润了。我递过去几张纸巾，张小锦执拗不接，只是不停地眨巴眼睛，希望眼帘把泪水吸干，睫毛就聚成几把纤巧的小刷子。

看来张小锦家充满了矛盾和危机，她妈妈的朋友也许正是罪魁祸首。我说："小锦，是妈妈的朋友让你们家庭变得不幸福了？"

张小锦一个劲儿地点头："正是！"

我说："妈妈的坏朋友具体是个怎样的人？"

张小锦突然有点踌躇，说："其实这人也不算太坏，逢年过节都会给我买礼物，是我妈的闺蜜。"

晕！我一直以为妈妈的朋友是个男人，甚至怀疑他就是破坏张小锦家的第三者。现在才知道，朋友是个女的！有一瞬间，闪过张小锦的妈妈是不是个双性恋的念头。要不然，怎么两个女人之间的关系会引发张小锦这样大的恼怒！

咨询师的脑海就像一台高速运转的电子计算机，来访者的任何一句谈话，都会在咨询师脑海中引发涟漪。一千种可能性像漂流瓶在波涛中起伏，你不知道哪一只瓶内藏着来访者心中的魔兽。也许你以为是症结所在，穷追不舍，紧紧跟踪，结果不过是一朵七彩泡沫。也许你忽视的只言片语，却潜藏着最重要的破解全局的咒语。这一次，我的方向差了。

我想起了老师的教导："你不能以自己的主观猜测代替事实的真相。你永远不能跑到来访者的前面去，你只能跟随……跟随……还是跟随。"

我调整了心态，对张小锦说："你妈妈和女友之间的关系，让你嫉妒。"

张小锦不解地重复："嫉妒？我好像没有想到这一点。"

Zhang Xiaojin concurred: "Exactly! But that 'friend' of my mom's is always whispering to her, urging her to divorce, plotting to tear our family apart!" Saying, her eyes misted over. I handed her some tissue, but she brushed them aside and kept batting her eyelids, trying to shake off the clinging tear drops.

It seemed that Xiaojin's family was full of tension. Perhaps her mom's friend was at the bottom of the trouble. I asked: "Xiaojin, could this friend be at the bottom of your family's troubles?"

Xiaojin nodded vigorously, "Exactly!"

"What is this friend like, this troublemaker?"

Xiaojin shifted uneasily, saying, "Well, she's not *that* bad. She gives me presents for New Year.... She is my mum's buddy."

I had assumed that the friend was male, a seducer. And now it turns out that the buddy was a woman. It flashed across my mind whether this could be a bisexual relationship, in which case it would explain Xiaojin's resentment.

In fact, the mind of the counselor is like a computer, casual remarks by the visitor would set it spinning off in different directions. A thousand possibilities started spinning wildly in my head, like bottles floating on the waves. One cannot tell which of these bottles was holding the secret that is gnawing at the heart of this young girl. Sometimes, you are actually chasing a bubble when you think you have nailed down the problem. On the other hand, a casual word that you have overlooked could actually be the key. This time, I had taken a wrong direction. I recalled the admonition of my teacher: Do not rely on your own deduction in place of solid fact. Never ever leap ahead of the visitor seeking counsel. All you can do is to follow behind, always following behind.

I adjusted my mood, and said to Xiaojin: "You are jealous of the relationship between your mom and her girlfriend."

Xiaojin mumbled to herself: "Jealous? The idea had never occurred to me."

我说：“以前没想到不要紧，现在开始想也来得及。”

张小锦偏着脑袋想了一会儿说：“好吧，你说我嫉妒，我承认。人家都说女儿是妈妈的小棉袄，可我妈妈硬是把我当成了破大衣，心里有话都不跟我讲。”

我说：“你妈妈的心里话是什么呢？”

这一次，张小锦反常地沉默了，很久很久。如果我不是一个训练有素的心理师，也许我就睡着了。我等待着张小锦，我知道这些话对她一定非常重要，讲出口又非常困难。

终于啊终于，张小锦说：“哼！他们都以为我不知道，他们合伙儿来骗我。我也愿意装出一副傻相，让他们以为我不知道。他们自以为知道一切，其实我在暗里比他们知道得更多！”

简直就是一个绕口令！我彻头彻尾被这个有着破碎长笛一样沙哑嗓音的女生弄糊涂了。我要澄清，在她的词典里，“他们”——是谁？

“是我爸爸、我妈妈，还有那个和我爸爸相好的女人。当然，还有我妈妈的闺蜜……”张小锦的话匣子终于打开了。

"Well," I said, "You can start thinking about it now."

Zhang Xiaojin held her head sideways, thinking. "Alright," she announced, "I admit I am jealous. There is a saying that 'daughters are like cotton-padded jackets to warm their mothers' hearts'. But somehow, my mother treats me like a ragged old coat, she never confides in me about anything."

I followed with the question: "What is the problem that she should have confided in you?"

Zhang Xiaojin remained silent for a long time. If not for my training as a professional counselor, I would have fallen asleep. I knew that the answer to my question will touch a raw nerve and that it is not easy for her to spell it out. Finally, Xiaojin muttered, "They think that I don't know anything, they try to keep me in the dark. I don't mind pretending. Let them think that I am still in the dark. Actually I know more than they think!"

Such a tongue twister! So many they's and them's, she got me all confused. The main thing is to clarify who are the people that she kept referring to as "they."

Xiaojin finally decided to get everything off her chest. "They are my father, my mother, the woman who is entangled with my father, and then there is my mother's buddy!"

原来，张小锦的爸爸有了外遇，和另外一个女子暧昧，被放学回来的张小锦撞见了。从此，张小锦见了爸爸不理不睬，爸爸反倒对张小锦格外好。张小锦决定不把这件事告诉妈妈，因为那样家就很可能破碎。张小锦知道那些父母离婚的同学基本上都很自卑。张小锦心想，只要妈妈不发现这件事，家庭就能保全。她一次又一次地帮着爸爸遮掩，让妈妈蒙在鼓里。然而，妈妈还是察觉到了某种蛛丝马迹，开始敏感而多疑。张小锦很怕出事，就故意胡闹，分散妈妈的注意力，实在没法子了就生病。无论妈妈多么在意爸爸的一举一动，只要张小锦一发烧，妈妈就把所有的注意力都放到了张小锦身上，无暇他顾，爸爸的危机就化解了。可爸爸不知悔改，变本加厉。张小锦就是再用十八般武艺转移妈妈的注意力，妈妈还是越来越接近真相了。妈妈对自己的好朋友痛哭一场和盘托出。这位闺蜜是个刚烈女子，疾恶如仇。她不断和妈妈分析爸爸的新动向，号召妈妈奋起抗击。妈妈很痛苦，和闺蜜无话不谈，最近已经到了商议如何去法院告道德败坏的爸爸，讨论分割财产和张小锦的归属……张小锦用大量的精力偷听她们的谈话，惊恐万分。好比外敌入侵，妈妈的闺蜜是主战派，张小锦是主和派。张小锦要维护家园，当务之急就是除掉闺蜜！她走投无路，不知道跟谁商量。跟同学不能说，要维持幸福家庭的假象；跟亲戚不能说，爸爸妈妈都是好面子的人，张小锦不愿亲人们知道家中正在爆发内乱；跟老师也不能说，她害怕老师从此把她归入需要特别关心爱护的群体。百般无奈的张小锦想到了心理医生，就把所有的私房钱都拿出来做了咨询费。

The fact was, Xiaojin's father had been carrying on an affair, and Xiaojin had bumped into them one day when she got home from school. From then on, her father pampered her every way he could, but Xiaojin ignored him. However, she decided not to tell her mother, as that might lead to the breakup of the family. Xiaojin assumed that as long as her mother was in the dark, the family will keep going. She knew from school that kids whose parents were divorced had low self-esteem. She figured that so long as her mom is kept in the dark, their family will be intact. Again and again, she had helped shield her dad from discovery. But somehow her mom began to be suspicious. So Xiaojin feigned illness and resorted to other tricks to distract her mom. Luckily for Xiaojin, no matter what is at stake, the moment she got ill, her mom would drop everything to take care of her. And thus her father's affair evaded detection. Unfortunately, her dad was a man who did not keep within limits. No matter how hard Xiaojin had tried to shield her mom, the poor woman could not help sensing that something was wrong. She confided tearfully to her buddy, and this buddy, happening to be the vindictive type, encouraged her mom to fight her husband tooth and nail, so much so that things had come to a head and the two women were on the verge of dragging Xiaojin's dad to court as a sample of moral depravity. They had gone so far as to discuss division of property and custody of the child!

Zhang Xiaojin had spent a lot of time eavesdropping on their conversation and she became alarmed. Her mom's buddy was all for confrontation, while her mom was inclined towards a reconciliation. For Xiaojin likewise, the top priority was to keep the family intact. To achieve this, she must get rid of the buddy! She did not know where to turn for help. Not to her school mates, as she had to keep up appearances. Not to relatives, as both her dad and mom must save face. Not to her teacher, or she might end up being handled gingerly as a problem child. As a last resort, she thought of psychological counseling, and took out all her savings for the counseling fee.

听完了这一切，我把张小锦抱在怀里，她像一只深秋冷雨后的蝴蝶，每一根发丝都在极细微地颤抖。不知道在这具小小的躯体里隐藏了多少苦恼与愤怒！她还是个孩子啊，却肩负起了成人世界的纷争，为了自己的家庭，咽下了多少委屈、辛酸的苦果！

许久后，我说："小锦，设想一个奇迹。假如你妈妈的闺蜜突然消失了，你们家就能平静吗？"

张小锦认真想了一会儿，说："可能会平静几天吧。但我妈妈已经起了疑心，她会穷追到底，我爸爸迟早得露馅。"

我说："这么说，闺蜜并不是事情的症结……"

张小锦是个聪明孩子，马上领悟过来，说："事情的根本是我爸妈自己！"

我说："你同意我请你的爸爸妈妈到这里来咱们一同讨论你们家的情况吗？"

张小锦害怕地抱着双肩说："他们会离婚吗？"

我说："不知道，咱们一块儿努力吧。只是有一条，这一次，你不能装作什么都不知道，你要把你所知道的一切和感受都说出来，包括你对父亲第三者的印象，还有你对闺蜜的看法。你要表达你对父母的期待和对一个完整的家的爱。"

张小锦说："天哪！在爸爸妈妈眼里，我一直是个善解人意的乖乖女，这下子，我岂不是变成了刺探情报两面三刀的小克格勃？！不干！不干！"

我说："这是否比你失去爸爸妈妈和家庭瓦解更可怕？"

After hearing through Xiaojin's recital, I held her close to my breast. She was so fragile, like a butterfly after a spell of rain, every hair on her head trembling in the cold. Who can tell how much anger and anxiety were stored in that fragile frame? Still a child but burdened with the strife of the adult world! Who can tell how much she had suffered on account of her family!

After a while, I said: "Xiaojin, suppose there is a miracle and your mom's buddy disappeared. Will there be peace in your family?"

Xiaojin thought for a moment and said: "Perhaps peace for a little while. But my mom is already alerted. She will not give up until she ferrets out the truth. My dad is bound to be caught."

"In that case, the buddy is not the critical element in this affair..."

Xiaojin was quick to grasp my point. "The critical problem is between my dad and my mom!"

"Would you approve if I invite your mom and dad to come over and we sit down and talk through their problem?"

Arms crossed over her chest, Xiaojin clasped a shoulder with each hand, hugging herself, as if frightened at the prospect.

She asked: "Will they divorce?"

I said: "I don't know. Let us do our best to help. But this time you must not pretend to be in the dark. You must speak up. Tell them all that you know. Speak up about your feelings, including your impression from the woman that your father is involved with. Speak up about your impression of your mom's buddy. Let your parents know what you expect from them. Let them know that you want your family intact."

"Good heavens! I have always been a good daughter to my parents. I am not going to be seen as a spy, a little KGB. Oh no!"

"Is it worse than a broken home and the loss of your parents?"

张小锦捂着眼睛说：“好吧。我知道什么事最可怕。”

我们和张小锦的爸爸妈妈取得了联系，他们一同来到咨询室。经过多次的家庭讨论，这其中有很多激战和眼泪，张小锦的爸爸终于决定珍惜家庭，和第三者一刀两断。妈妈也说看在小锦一番苦心上，给爸爸一个痛改前非的机会。

结束最后一次咨询，张小锦离开的时候悄悄地对我说：“现在，我也有了一个闺蜜，给我出了个好主意。”

我说：“谁呀？”她说：“就是你啊！”

Zhang Xiaojin covered her eyes with both hands. “Alright, go ahead. I know which is worse.”

We got into touch with Zhang Xiaojin’s parents and they showed up. After a lot of talk, including tears and mutual recriminations, Xiaojin’s father decided to break off the clandestine relationship and stick to the family. Xiaojin’s mom appreciated all that her daughter had done to keep the family together and decided to give her husband a chance.

Wrapping up our last consultation, Zhang Xiaojin confided:

“I have a buddy who gives me good advice.”

“Who is that?”

“You, of course!”

谁是你的闺蜜

Best Buddies

生命是单独的个体，

无论怎样的血肉交融，

我们必须独自面对

世界的风雨。

藏獒与虎皮鹦鹉

一天，咨询室来了一位少女。相貌平平，身材中等，神情也很寡淡，没有热情，也没有强烈的拒绝或是好奇。她穿着一身浑浊的白色纯棉衣服，带着很多口袋和皱褶。让人不由得想起一块微潮的抹布，既不可能很快燃烧，也拧不出水来。

陪同她来的是她的母亲，一位富态而考究的中年妇女。当着妈妈的面，这女孩是委顿和沉默的。妈妈被留在门外之后，她坐在沙发上，脸孔始终向着窗户的方向，好像随时准备站起身来钻窗而去。

她说自己叫桑如，是被妈妈强逼着来看心理医生的。

我问她："你多大了？"（其实，我手旁就放着她的登记表，那上面写着她的年龄。但我愿意请她亲口再次说出自己的岁数，这样有利于提醒她对自己负起责任。）

"我上高一，16岁半了。"桑如回答说。

我点点头，说"对于一个16岁的人来说，是可以自己决定要不要看心理医生的。如果你非常不愿意，我觉得你可以走了。至于你母亲那里，由我来和她解释。我会说，我们要尊重一个16岁的青年人对自己的判断。如果她觉得自己不需要心理医生的帮助，自己完全可以解决问题，那很好。现在，如果你想离开，门在那边。"

Between the Mastiff and the Parakeet

A young girl walked into my office one day. Medium height, average looks, face expressionless, showing neither interest nor aversion to the surroundings, no curiosity whatever. She was in a cotton dress of an indeterminate pale color with many folds, reminding me of a bedraggled piece of wet cloth which could hardly serve any useful purpose.

With the girl was her mother, a stout, well-dressed, middle-aged woman, obviously well-to-do. In the presence of her mother, the girl remained subdued, silent. As she and I entered the consulting room without her mother, the girl strode to the sofa and sat down, eyes fixed on the window, as if she could fly out at any moment.

The girl said that her name was Sangru and that her mother had made her come for counseling.

I asked for her age. It was set down in the register, of course, but I wanted her to hear herself saying it. To remind her that she was responsible for herself.

"I am sixteen-and-a-half years old. I am in first year senior high."

I nodded and told her: "As someone in senior high, you have the right to decide whether you wish to see a counseling psychologist. If you would rather not, you can get up and go this minute. I will do the explaining to your mother. I will tell her that we must respect the decision of a sixteen-year-old about her own condition. I will tell your mother that if her daughter thinks she can deal with her own problems and do not need counseling, then she should be let go. Now, the door is there, you may leave."

这个建议的确不是我个人的心血来潮。在美国访问的时候，美方心理医生告诉我，对于10岁以上的少年，他们都会征询来访者的意见。如果该少年强烈地拒绝接受心理医生的辅导，便不会强迫他们。这是很有道理的——心理医生是助人自助的工作，如果某人拒绝帮助，那么你就是有再大的热情，也是枉然。

桑如对我的回答显得很意外，甚至有点儿不知所措。她反问道："这是真的吗？"

我说："当然是真的了。"

她愣了半天，好像对突然获得的自由很有点儿不习惯，站起身来，又坐下了，说："如果您真让我决定，那我就不走了。我觉得您和别的成年人不一样，居然让我自己作决定。"

我笑起来，说："你指的别的成年人，是谁呢？"

桑如用手指点点门外，说："比如说我的爸爸妈妈啊。他们不让我做这个，不让我做那个，什么都要听他们的。连我用的卫生巾买什么牌子，都得我妈妈说了算。真是太烦了！"

我深深地点点头，表示明白了她的心境。我说："谢谢你这样信任我。既然你愿意接纳我的咨询，那么，咱们今天要讨论什么问题呢？"

桑如说："我妈妈本来是想让我和您谈谈人际关系的事儿。我不愿意和别的同学交往，自己也很苦恼，我不知道如何和别人相处。可我今天不想谈这个问题，我希望你能帮我决定一下，到底养个什么宠物好。"说完，桑如充满期待地看着我。

I was not speaking out of an impulse. When I was in the US, an American counseling psychologist told me that they respect the decision of young people over ten years old as to whether they need counseling. Nothing will ever be forced on them if they were unwilling. This makes sense—psychological counseling is a means of helping people help themselves. If people refuse such help, nothing that you force on them will ever work. .

Sangru seemed surprised at my response and asked if I meant what I said.

"Certainly," I told her, "I mean what I say.".

For a moment, she was dumbstruck, shocked by this newly acquired freedom. She shot up, stood still for a moment and sat down again. "If you really mean that I can make the decision, I will stay. You are so unlike other adults. You actually let me make the decision!"

I laughed. "Whom do you mean by *other adults*?"

Sangru pointed to the door. "Them, my mom and dad. It's always you mustn't do this, or you mustn't do that. It's always what they want. Even my sanitation pads have to be a brand that my mom approves. It's such a drag."

I nodded vigorously, signaling that I understood."Thank you for trusting me," I said. "If you accept my counseling, where should we start?"

"My mother had originally planned for me to talk about socializing. I am not interested in getting together with my classmates. But then I am lonely. I don't know how to get along with people. I don't want to talk about this now. For starters, I wish you to help me decide what pet I should get—a Tibetan mastiff, or a parakeet?" She looked at me expectantly.

现在，轮到我稍稍忐忑了。养宠物？这对我真是一个新问题，第一个念头是——我也不是个兽医，怎么会知道哪种宠物好？好在我很快整理好了自己的思绪，捕捉到这个问题的核心——不在于桑如到底养一只什么样的宠物，而在于她为什么生出这个念头。

不过目前我不能走得太快，只能跟随在桑如后面。我说："好啊，我很愿意帮你出出主意。你打算养的宠物备选名单是什么？"

桑如一下子变得兴致勃勃，说："我最想养的是藏獒。"

我吓得差点儿没从沙发上跌落下来。我在西藏当过兵，见过这种凶悍无比的犬类。它们高大威猛，极端吃苦耐劳和忠于主人，为了保护羊群，甚至可以和群狼搏斗。藏獒当然是值得尊敬的，但面前这样一个清瘦女孩，要把藏獒当宠物养，恐怕并不相宜。

我说："藏獒需要很大的活动场地，属于旷野。再说它们是大型犬，城里不让养的。价钱也很贵，我记得看过一篇报道，一只好的藏獒要很多钱呢。"

桑如立刻说："哈巴狗如何？"

藏獒和哈巴狗，在性格上实在是南辕北辙，桑如居然这么快就改弦易辙，这让我对她养宠物的初衷产生了更大的纳闷儿。我说："你为什么又改养哈巴狗呢？"

桑如说："我喜欢藏獒的忠诚，但城里不让养，我也没办法。哈巴狗对人非常友善，而且，你让它干什么，它就干什么。整天围着你的裤腿转，可会讨好人了。"

Now it was my turn to be hesitant. Choosing a pet? It is something new on my plate. My first reaction had been: "I am not a vet. How should I know what is best to choose for a pet?" Then I collected my thoughts, and got to the heart of the matter—not which kind of pet to choose, but why she seized on such an idea in the first place! I did not want to jump ahead of her line of thought. I asked, "Good, I would like to help you make a decision. What kind of pet do you have in mind?"

Sangru was elated, "I would like to keep a Tibetan mastiff," she said.

I was so shocked I nearly fell off my seat. I had spent years in the army in Tibet and had seen those mastiffs. They are huge and powerful, fiercely loyal to their masters and totally reliable in protecting the family flock, even fighting roving bands of wolves. But I do not think them suitable as pets for a young girl. I began by saying that those hounds are to be respected, but that they are not easily domesticated and need vast spaces to move about. Besides, the city precincts are off limits to them. They would be too much for a slim young girl to handle."

Sangru immediately said: "Then I will raise a poodle!"

From a Tibetan mastiff to a poodle, that was really a fast move. I was curious—why did she want a pet in the first place?

She said "I admire the loyalty of the Tibetan mastiff. But since they are off limits in the city, I will shift to poodles. Poodles nestle at your feet and obey commands. They are ever so cute."

原来是这样！我说："哈巴狗倒是可以养的。只是，你会每天喂它吗？你每个星期给它洗澡吗？你愿意清扫它的狗窝，清理它的粪便吗？还有每天都要带着它到外面去撒欢？如果用行话来说，就是遛狗。狗还要按时打预防针，如果你的哈巴狗把你自己或是别人给咬伤了，就要马上到医院注射狂犬疫苗，并且不是一次就能完成，要好几次……"

桑如吓得吐了吐舌头，说："呦！这么复杂！"

我说："还有更多的事情等着你。如果它病了，你要抱着它到宠物诊所看病，如果它需要手术，你要守候在它身边……"

桑如听到这里，连连摆手说："天啊！这么麻烦！那我不养狗了，我改养一只猫，这就简单了吧？"

我说："恐怕也不像你想的那样简单。首先，猫像狗一样，也要大小便，这样你收拾它们排泄物的工作并不能免掉。猫也要洗澡，也要到外面去玩耍。它们在外面的时候，你不知道它们会捕获什么猎物，也许是虫子，也许是小鸟，这是猫的天性，你不能阻止它们，它们也许会带回来一些病菌。春天是猫繁殖的季节，它们会大声叫，如果你不想让它们叫，就要给猫做手术。而且，猫为了磨炼它们的爪子，一般都会撕纸，这也是它们的天性……"

桑如惊叫起来说："我的作业本！我的书！如果猫不管不顾地撕坏了它们，我要重写一遍吗？天哪，我不养猫了，我养一对小鹦鹉成不成呢？"

我说："当然可以啦！"

桑如说："我会训练让它们说话，有它们和我做伴，我就不会寂寞了。"

我心中一动，明白了症结就在这里，但此刻不是点破的好时机。顺着桑如的思路走，我说："是那种翠绿的虎皮鹦鹉吗？"

So that was it, I thought to myself. I said, "Poodles are easier to handle. But are you ready to feed them every day and bath them every week? Will you clean up their box and dispose of their droppings? Will you take them out for their daily walk? Moreover, you must take them to the vet for shots. If you or someone is bitten by your doggie, you must take the person straight to a doctor for treatment, and more than once..."

Sangru rolled her eyes, "So much trouble!"

"That is not all," I added, "if your doggie gets sick, you must hold it in your arms and take it to a vet. If it needs an operation, you must stay with it..."

Sangru raised her hand, "Heavens, so much trouble. Forget dogs. I will get a cat, much easier to handle. Right?"

"Not as easy as you think. Like dogs, cats need to pooh and urinate. They also need to bathe and to go out for strolls. So that part of the work remains the same. But with cats, you never know what they will bring back from outside—dead birds and insects, and that means germs. Come spring, they caterwaul all night. If you don't like the sound, you must go to the vet and get the cat castrated. Besides, cats are always tearing and clawing with their nails, tearing up pieces of paper, it is part of their nature... "

Sangru gave a little scream: "Oh no, my books, my homework! I have to rewrite my homework if it is torn! Oh no! No cats! What about a parakeet?"

I said, "Sure, you can keep a parakeet."

Sangru added, "I will train it to talk. It will keep me company and I won't be lonely."

My heart missed a beat. So that was the heart of the matter.

But now was not the right moment to spell it out.

Following Sangru's line of thought, I asked: "Is it the jade green kind called 'budgerigar'?"

桑如说：“对啊。就是有绿色蓝色黄色羽毛的小鹦鹉，好像穿着丝绸的外套，闪闪发光。”

我字斟句酌地说：“鹦鹉的羽毛是需要经常打扫的，不然飘落在地上变成粉尘，很容易传播疾病。我记得有一种传染病，就叫‘鹦鹉热’。另外，据我所知，这种花色灿烂的小鹦鹉并不会说话，会说话的那种叫作鹩哥。就算有个别极其聪明的小鹦鹉，你训练到它能说话了，估计也是非常简单的‘你好’之类，并不如你想象的那样可以陪着你聊天。就算最棒的鹩哥，能说的话也很有限，它们只是模仿，并不会动脑子。”

桑如生起气来，打断我的话：“您这也不让养，那也不让养，处处给我设障碍！”

我说：“桑如，我并不是不让你养宠物，我没有这个权力。只是，我认为每一个预备养宠物的人，都要事先搞清楚宠物的脾气，知道它们的习性，觉得自己能够担当得起，再来和宠物相处。如果因为自己有很多问题解决不了，以为养了宠物就可以逃避现实，那不但是对自己不负责任，也是对宠物的不负责任。毕竟，人间的问题，只有在人间解决。”

桑如听了我的话，许久不做声，看得出非常沮丧。

我轻轻地问：“桑如，你能告诉我你为什么要养宠物吗？”

就这样一句普通的问话，却让桑如哭泣起来。她说：“我太孤独了。同学们都不愿意和我玩儿，她们说我是一个没意思的人。我不知道如何交往，就想，哼！你们不理我，我也不理你们，我要和小动物一起生活。只有它们不会嫌弃我，不会嘲笑我，会很忠于我。我让它们干什么，它们就会干什么，绝不会背后议论我……它们是多么安全可靠的朋友啊……”

Sangru said: "Right, the miniature type with blue, green and yellow feathers, as if wearing a glistening multi-colored jacket."

I stated, word by word: "A parrot's feathers must be frequently dusted, otherwise they fall off and are crunched to dust underfoot and spread disease. It is called 'parrot fever.' Besides, this kind of small size parrots do not speak. Even if a smart little parrot learns so say 'hello' after a lot of training, still their speech is purely mechanical, never a real conversation as you imagine. And even if it does speak, its 'vocabulary' is very limited, and purely repetitive."

Sangru became annoyed, and cut me short saying, "I can't do this and I can't do that! Obstructions every way I turn!"

I said: "Sangru, I am not forbidding you to keep a pet. I do not have the right. But you must find out all that is involved and make sure that you are up to the job. If you want to escape your problems by keeping a pet, it is being irresponsible, both to you yourself and to the pet. After all, human problems must be resolved by human beings in the human world." Sangru heard me out and remained silent for a long time, looking very crestfallen.

"Sangru, would you like to tell me why you want a pet?"

Sangru burst out crying at this very simple question. "It is so lonely!" she murmured, "At school no one wants to be my friend, they say I am boring. I don't know how to approach people. I decided that since they ignore me, I will ignore them back. I was going to have fun with a pet—my pet will not ignore me or taunt me. My pet will be obedient and will do whatever I say and not gossip about me behind my back. I will feel safe and comfortable with such friends..."

原来，桑如想养宠物的初衷，是为了解决自己的人际关系问题。看来，桑如母亲的判断并没有出错，只是，桑如的母亲也许不知道，女儿的怯懦无趣，正是和母亲对桑如的过度保护有关。在母亲眼里，桑如永远是长不大的小女孩儿，一切都要由母亲做主。这一切，让桑如不曾学会如何同小伙伴们相处，到了青春逆反期，就越发孤独。桑如很苦恼，她要为自己寻找一个突破口，找到温暖与信任，找到安全与友爱，于是，她只有求助于宠物。

我全神贯注地倾听桑如的痛苦，听她断断续续地说被同学奚落和孤立……第一次的咨询时间很快就过去了。临走的时候，桑如怯生生地说："我下周还可以来看您吗？"

我说："当然可以了。不过，这要你自觉自愿，不要妈妈陪着。"

桑如破涕为笑，说："当然是我自愿的。"

私下里，我同桑如的妈妈谈了一次。当然，我并没有把和桑如的具体谈话内容告诉她妈妈，只是希望桑如妈妈能正视女儿已经16岁了这个现实，该放手的时候就要大胆放手。桑妈妈答应了。

经过若干次的咨询以后，桑如渐渐灵动起来，她开始学会和同学们友好相处，甚至还准备了一些小笑话，打算在春游的时候讲给同学们听。咨询时，她说："我先试着把笑话讲给您听听，如果您笑了，我就敢对着同学们讲了。"

我认真地听完了桑如的笑话，开心地笑了。说实话，不是桑如的笑话有多么可笑，实在是我看到了她的成长，充满欢愉。

桑如后来说过一句让我长久不能忘怀的话，她说——人最好的宠物，其实就是自己。

Now it is clear that Sangru wanted to keep a pet to avert her problem with relationships. Obviously Sangru's mother knew that her daughter had a problem with socializing. But she didn't see her daughter's awkwardness and timidity, which was actually due to her own over-protectiveness ever since Sangru had been a little girl. In the mother's eyes, Sangru is a child who never grew up and mom had to take care of everything. During the rebellious interval of puberty, things went from bad to worse. In her lonely unhappiness, Sangru had wanted to find a way out for herself. She was looking for warmth and trust, for friendship and a sense of security. So she seized on the idea of a pet.

I listened with great concentration to Sangru's tale of her woes, of how she had been isolated and taunted by her classmates.

The hour of counseling was soon up.

As she was leaving, Sangru asked hesitantly: "May I come again next week?"

"Of course," I answered, "but it must be your own decision. Not dragged over by your mom."

Sangru smiled through her tears, "Of course I will be coming on my own decision."

Later, I had a private conversation with Sangru's mom. I did not reveal the substance of my conversation with Sangru. But it was finally driven into her head that her daughter was sixteen years old and that she as the mom should let go. She promised to do so.

After several successive sessions of counseling, Sangru became more relaxed and started to move easily among her schoolmates. She even stocked up some jokes against the arrival of the New Year celebrations. She said she will try them on me first before she uses them on her school friends. When Sangru told her jokes, I did have a hearty laugh—less at the jokes than at Sangru's rapid growth.

Sangru told me something that I will always remember:

"Your best pet is you yourself."

红与黑的少女

来访者进门的时候，带来了一股寒气，虽然正是夏末秋初的日子，天气还很炎热。

女孩，十七八岁的样子，浑身上下只有两种颜色——红与黑。这两种美丽的颜色，在她身上搭配起来，却显得恐怖。黑色的上衣黑色的裙，黑色的鞋子黑色的袜，仿佛一滴细长的墨汁洇开，连空气也被染黑。苍黄的脸上有两团夸张的胭脂，嘴唇红得仿佛渗出血珠，该黑的地方却不黑，头发干涩枯黄，全无这个年纪女孩青丝应有的乌泽。眼珠也是昏黄的，裹着血丝。

“我等了您很久……很久……”她低声说自己的名字叫飞茹。

我歉意地点点头，因为预约人多，很多人从春排到了秋。我说：“对不起。”飞茹说：“没有什么对不起的，这个世界上对不起我的人太多了，你这算什么呢！”

飞茹是一个敏感而倔强的女生，我们开始了谈话。她说：“你看到过我这样的女孩吗？”

The Girl in Red and Black

A cold wave followed in her wake as she swept into the room, while outdoors the heat was still lingering through the last days of summer.

Sweet seventeen, or eighteen years old, I would say, with only two colors to her outfit, red and black. She looked scary, wrapped in black from top to toe. Black shirt over black skirt, black shoes over black stockings—she looked like a monstrous blob of black ink seeping downward into a giant smear, the blackness permeating the room. Her lipstick was so heavily applied and her cheeks so generously rouged that she seemed to be bleeding. Her hair, however, which one expected to be a shiny black, was on the contrary dry and seared, totally unlike the glossy hair for girls of her age. Her eyes too, were glazed, her eyeballs laced with wisps of blood.

She murmured: "I have been waiting so long...so long," as she gave me her name—Feiru.

I nodded. Indeed there were so many applicants lined up—one applies in spring and gets an appointment for the fall. I apologized. Feiru said no need to apologize. She added that there were so many people in the world to whom she owed an apology.

Feiru seemed tough, but sensitive at the same time. She started the conversation by asking me: "Have you ever seen a girl like me?"

我一时不知如何回答好，就说：“没有。每一个人都是特殊的，所以，我从来没有看到过两个思想上完全相同的人，就算是双胞胎，也不一样。”

这话基本上是无懈可击的，但飞茹不满意，说：“我指的不是思想上，我知道这个世界上绝没有和我一样遭遇的女孩。打扮上，纯黑的。”

我老老实实地回答：“我见过浑身上下都穿黑衣服的女孩。通常她们都是很酷的。”

飞茹说：“我跟她们不一样。她们多是在装酷，我是真的……残酷。”说到这里，她深深地低下了头。

我陷入了困惑。谈话进行了半天，我还不知道她是为什么而来。主动权似乎一直掌握在飞茹手里，让人跟着她的情绪打转。我赶快调整心态，回到自己内心的澄静中去。这女孩子似乎有种魔力，让人不由自主地关切她，好像她的全身都散发着一个信息——“救救我！”可她又被一种顽强的自尊包裹着，如玻璃般脆弱。

我问她：“你等了我这么久，为了什么？”

飞茹说：“为了找一个人看我跳舞。我不知道找谁，我在这个大千世界上找了很久，最后我选中了你。”

我几乎怀疑这个女生的精神是否正常，要知道，付了咨询费，只是为了找一个人看跳舞，匪夷所思。再加上心理咨询室实在也不是一个表演舞蹈的好地方，窄小，到处都是沙发腿，真要旋转起来，会碰得鼻青脸肿。我当过多年的临床医生，判断她并非精神病患者，而是在内心淤积着强大的苦闷。

Caught unprepared, I murmured "No." And so what if I haven't, I thought to myself. Actually people are each unique in their own way. I have never seen two people totally alike in personality—not even twins. It is a truth *universally acknowledged* that no two people could be exactly alike. But that was not the point that Feiru was trying to make.

She said, "I don't mean two people being the same in the way that you assume. But I know for sure that there is not another girl in the whole world who had to go through what I had to..." Does she mean having to be dressed in black from top to toe? I wondered.

I said: "Of course I have seen girls dressed in black from top to toe. It is considered *cool*, I believe."

"I am different. They are pretending to be cool. As for me, I am the real thing, I am *cruel*." Saying which, she dropped her eyes.

I was perplexed. She had steered the conversation whichever way she wanted, but after hearing out her ramblings, I still did not know why she was here. I collected myself and tried to keep my cool. Somehow the girl did have a hold over me, appealing to me with her whole being, as if saying: "Please save me!" On the other hand, however, she seemed to be taking refuge inside a glass cage, wrapped within her own ego.

"Why were you waiting this long to see me?"

She said: "I must find someone to watch me dance. I have been looking and looking for someone in the whole wide world. Finally I decided to ask you."

For a moment, I was unsure whether she was in her right mind—to have paid the fee and waited so long—just for someone to watch her dance! Besides, the psychological counseling office was not at all conducive to dancing. It was small and cramped with a sofa. If one started to waltz around, one would be knocked about by the legs of the sofa. I decided that she was not really a mental case weighed down by unhappiness.

我说："你是个专业的舞蹈演员吗？"

飞茹说："不是。"

我又说："但这个表演对你来说，非常重要。为了这个表演，你等了很久很久。"

飞茹频频点头："我和很多人说过我要找到看我表演的人，他们都以为我是在说胡话，甚至怀疑我不正常。我没有病，甚至可以说是很坚强。要是一般人遇到我那样的遭遇，不疯了才怪呢！"

我迅速地搜索记忆，当一个临床心理医生，记性要好。刚才在谈到自己的时候，她用了一个词，叫作"残酷"，很少有正当花季的女生这样形容自己，在她一身黑色的包装之下，隐藏着怎样的深渊和惨烈？现在又说到"疯了"，到底发生了什么？

贸然追问，肯定是不明智的，不能跨越到来访者前面去，需要耐心地追随。照目前这种情况，我觉得最好的方法是尊重飞茹的选择：看她跳舞。

我说："谢谢你让我看舞蹈。需要很大的地方吗？我们可以把沙发搬开。"

飞茹打量着四周，说："把沙发靠边，茶几推到窗子下面，地方就差不多够用了。"

于是我们两个嗨哟嗨哟地干起活来，木质沙发腿在地板上摩擦出粗糙的声音，我猜外面的工作人员一定从门扇上的"猫眼"镜向里面窥视着。诊所有规定，如果心理咨询室内有异常响动，他人要随时注意观察，以免发生意外。趁着飞茹埋头搬茶几的空子，我扭头对门扇做了一个微笑的表情，表示一切尚好，不必紧张。虽然看不到门扇后面的人影，但我知道他们一定不放心地研究着，不知道我到底要干什么。其实，我也不知道下面会发生什么事情，只是相信飞茹会带领着我，一步步潜入到她封闭已久的内心。

I asked her: “Are you trained as a professional dancer?”

“No.”

“Is this dance very important to you—something that you have been looking forward to?”

Feiru nodded. “I have asked around for someone to watch me dance, but nobody took me seriously; some even suspected that I was gaga. I am perfectly healthy, actually very strong. Anybody who had experienced what I had to bear would have gone crazy.”

I searched my memory—a psychological counselor needs a good memory. She had used the word “cruel” when referring to herself; young girls in the flower of youth rarely see themselves as such. I wondered what unmentionable darkness was hidden beneath that layer of black wrapping to make her refer to herself as cruel.

Of course it would be foolish and futile to pursue her with questions. The counselor’s job is to follow the track of her patient, not jump ahead. The best thing I could do at the moment, I decided, was to respect Feiru’s wishes and watch her dance. I thanked her for inviting me to watch her dance and offered to get rid of the sofa in the room to make more space. Feiru looked around and suggested that we push the sofa back against the wall and move the flower pot to the window shelf. The two of us set to work pushing and shoving, making a lot of noise. I knew that the staff outside would be checking on me through the peephole in the door. I lifted my head to give them an assuring smile. Though I could not see them through the peephole, I knew that they were there, wondering what I was up to. I myself was in the dark. But by then I was sure that ultimately Feiru would lead me into her inner self, which so far has been locked.

场地收拾出来了，诸物靠边，地中央腾出一块不小的地方，飞茹只要不跳出芭蕾舞中“倒踢紫金冠”那样的高难动作，应该不会磕着碰着了。我说：“飞茹，可以开始了吗？”

飞茹说：“行了。地方够用了。”她突然变得羞涩起来，好像一个非常幼小的孩子，难为情地说：“你真的愿意看我跳舞吗？”

我非常认真地向她保证：“真的，非常愿意。”

她用裹满红丝的眼珠盯着我说：“你说的是真话吗？”

我也毫不退缩地直视着她说：“是真话。”

飞茹说：“好吧。那我就开始跳了。”

一团乌云开始旋转，所到之处，如同乌黑的柏油倾泻在地，沉重黏腻。说实话，她跳得并不好，一点也不轻盈，也不优美，甚至是笨拙和僵硬的，但我一直目不转睛地看着，我知道这不是纯粹的艺术欣赏，而是一个痛苦的灵魂在用特殊的方式倾诉。

飞茹疲倦了，动作变得踉跄和挣扎，我想要搀扶她，被她拒绝。不知过了多久，她虚弱地跌倒在沙发上，满头大汗。我从窗台下的茶几上找到纸巾盒，抽出一大把纸巾让她擦汗。

待飞茹满头的汗水渐渐消散，这一次的治疗时到了结束的时候。飞茹说：“谢谢你看我跳舞。我好像松快一些了。”

飞茹离开之后，工作人员对我说：“听到心理室里乱七八糟地响，我们都闹不清发生了什么事，以为打起来了。”

我说：“治疗在进展中，放心好了。”

We finally managed to create a decent space within my consulting room and were ready for the dance. Feiru suddenly turned shy, asking whether I really wanted to see her dance. I assured her that I meant what I said. She turned to face me, asking if I was telling the truth. I looked her in the eye, and assured her that I did indeed. Finally Feiru was ready to begin.

Feiru's dancing was like a revolving black cloud heavily spread out over the floor. To be honest, her dancing was not much to speak of, being awkward and heavy. But I watched her without batting an eyelid. For me it was not an aesthetic experience. I was watching a tortured soul struggling to express itself in a tortuous fashion.

Gradually Feiru was tired and struggled to keep up the movements. I stretched out a helping hand, but she pushed me away. Finally she collapsed on the sofa, sweat streaming down her face. I offered her a wipe.

By the time Feiru had fixed herself up, the consultation was over and she was ready to go. As she left, Feiru said that she felt lighter at heart, now that someone had watched her dance. I assured her that the counseling was working and that she was doing fine. With Feiru gone, I assured my staff that things were moving forward and that they had nothing to worry about.

到了第二周规定的时间，飞茹又来了。这一次，工作人员提前就把沙发腾开了，飞茹有点意外，但看得出她有点高兴。很快她就开始新的舞蹈，跳得非常投入，整个身体好像就在这舞蹈中渐渐苏醒，手脚的配合慢慢协调起来，脸上的肌肉也不再那样僵直，有了一丝丝微笑的模样。也许，那还不能算作是微笑，只能说是有了一丁点的亮色，让人心里稍安。

每次飞茹都会准时来，在地中央跳舞。我要做的就是在一旁看她旋转，不敢有片刻的松懈。虽然我还猜不透她为什么要像穿上了魔鞋一样跳个不停，但是，我不能性急。现在，看飞茹跳舞，就是一切。

若干次之后，飞茹的舞姿有了进步，她却不再专心一意地跳舞了，说："你能抱抱我吗？"

我说："这对你非常重要吗？"

她紧张地说："您不愿意吗？"

我说："没有。我只是好奇。"

飞茹说："因为从来没有人抱过我。"

我半信半疑，心想就算飞茹如此阴郁，年岁还小，没有男朋友拥抱过她，但父母总是抱过吧？亲戚总是抱过吧？女友总是抱过吧？当我和她拥抱的时候，才相信她说的是真话。飞茹完全不会拥抱，她的重心向后仰着，好像时刻在逃避什么。身体仿佛一副棺材板，没有任何热度。我从心里涌出痛惜之情，不知道在这具小小的单薄身体中，隐藏着怎样的冰冷。我轻轻地拍打着她，如同拍打一个婴儿。她的身体一点点地暖和起来，柔软起来，变得像树叶一样可以随风飘曳了。

Feiru showed up promptly for her next appointment. By the time she arrived, the staff had already moved the sofa to make place for her dancing. Feiru was surprised, and obviously pleased. She started on a new dance. Her whole body seemed to be revived, the movement of her limbs were coordinated, her features lighted up by the hint of a smile. Perhaps the smile was still in the making, but it had a radiance which set my heart at ease.

In the days that followed, Feiru showed up for all her appointments. She danced in the middle of the consulting room, and my job was to watch her closely, without batting an eyelid. I was still in the dark about why she had to dance—dancing as if she was wearing a pair of fairy slippers and could never stop. For the moment, I could not afford to worry about it. Watching Feiru dance was top priority.

After an interval, when Feiru had made progress in her dancing, she seemed to lose interest.

One day she asked me: "Will you give me a hug?"

"Is it important to you?"

"Would you rather not?" she asked tensely.

"No, that's not what I meant! I was just curious."

She said: "Nobody has ever hugged me, *never*!"

I was a bit skeptical. Yes, she may be too young to have been embraced by a boyfriend, but what about parents? Relatives? Girl friends? But when we hugged as she had requested, I realized that she was telling the truth. She was so cold and awkward, bending backwards stiffly—it was obvious she had never hugged or been hugged in her life. I felt sorry for her. Who knows how much coldness had been cast upon her fragile little frame! I patted her softly as if cuddling a baby. Her body slowly warmed up and relaxed as if it could float away like a leaf in the wind.

下一次飞茹到来的时候，看到挤在墙角处的沙发，平静地说："您和我一道把它们复位吧。我不再跳舞了，也不再拥抱了。这一次，我要把我的故事告诉你。"

那真是一个极其可怕的故事。飞茹的爸爸妈妈一直不和，妈妈和别的男人好，被爸爸发现了。飞茹的爸爸是一个很内向的男子，他报复的手段就是隐忍。飞茹从小就感觉到家里的气氛不正常，可她不知道这是为了什么，总以为是自己不乖，就拼命讨爸爸妈妈欢喜。学校组织舞蹈表演，选上了飞茹，她高兴地告诉爸爸妈妈，六一到学校看她跳舞，爸爸妈妈都答应了。过节那天，老师用胭脂给她涂了两个红蛋蛋，在她的嘴上抹了口红。当她兴高采烈回家，打算一手一个地拉着爸爸妈妈看她演出的时候，见到的是两具穿着黑衣的尸体。爸爸在水里下了毒，骗妈妈喝下，看到妈妈死了后，再把剩下的毒水都喝了。

飞茹当场就昏过去了，被人救起后，变得很少说话。从那以后，她只穿黑色的衣服，在脸上涂红，还有鲜艳欲滴的口红。飞茹靠着一袭黑衣保持着和父母的精神联系和认同，她以这样的方式，既思念着父母，又抗议着被遗弃的命运。她未完成的愿望就是那一场精心准备的舞蹈，谁来欣赏？她无法挣扎而出，找不到自己存在的价值和重新生活的方向。

The next time Feiru showed up, she looked at the sofa pushed against the wall and said quietly: "Let's restore the sofa to where it belongs. I am not going to dance anymore, or ask for hugs. I am going to tell you my story."

It was a terrible story.

Feiru's parents did not get along. Her mom found someone else and her dad discovered the affair. He was introverted and kept things to himself. Feiru had sensed that something was wrong between her parents, but she could not lay her finger on the problem. She blamed herself and tried hard to be a good daughter to both parents. Once her school organized a show for June 1st Children's Day. Feiru had been chosen to perform a dance, and she had happily invited her parents to the show.

On that festive day, Feiru first went to school for rehearsal. She dressed up and her teacher touched up her cheeks with rouge and her lips with lipstick. After the rehearsal, Feiru ran home to fetch her parents. She ran into two corpses dressed in black. The fact was, her dad had poisoned the drinking water, watched her mom drink it down and die before his eyes. After making sure that she was dead, her dad finished off the rest of the poisoned water and followed his wife.

Feiru had fainted at the sight of the corpses of her parents, both dressed in black. After she was discovered and revived, Feiru stuck to black clothes, applied rouge to her cheeks and lipstick to her lips. She dressed exclusively in black to identify with her parents, at the same time putting on lipstick to protest against her orphaned state.

The only issue left hanging in the air was her dance—who was going to enjoy the dance that she had gone through such pains to practice? She could not lift herself out of the nightmare; she could not find her way to a new life.

对飞茹的治疗，是一个极为漫长的过程，我们共同走了很远的路。终于，飞茹换下了黑色的衣服，褪去了夸张的化妆，慢慢回归到正常的状态。

最后分别的时候到了，穿着清爽的牛仔裤和洁白衬衣的飞茹对我说：“那时候，每一次舞蹈和拥抱之后，我的身心都会有一点放松。我很佩服‘体会’这个词，身体里储藏着很多记忆，身体释放了，心灵也就慢慢松弛了。这一次，我和您就握手告别。”

Feiru's cure had been a long-drawn-out process, a rocky road that we walked together. Ultimately she grew out of the black clothes and went back to normal.

Feiru showed up in jeans and white top when we met to say goodbye.

She told me: "My whole body and soul would go through a relaxation whenever I danced." Now the dance would be followed by a hug for goodbye.

I like the word "go through." There are so many memories stored within us. As the baggage of memory is gradually released and flit away, our body and soul are relaxed.

Feiru and I shook hands for a last good-bye.

红与黑的少女

The Girl in Red and Black

即使是一块苦胆泡过的黄连，长久的咀嚼，也会使它的苦涩慢慢淡下去。

你究竟说了些什么

某天，一位朋友给我打电话，说："你到哪里去了？我找得你好苦啊！"因为是很好的朋友，我也和她开玩笑说："你是不是要请我吃饭啊？我欣然前往。"她着急地说："吃饭有什么难啊，事成之后，我一定大宴于你。只是我们现在要把事情做完，每拖延一天，损失就太大了。"

我听出她语气中的急迫，也就收敛起调侃，问道："到底出了什么事？"

她不容置疑地说："我要请你做心理咨询。"我松了一口气，说："你要做心理咨询，这很好啊。看来大家是越来越重视自己的心理健康了。只是我们是朋友关系，我不能给你做心理咨询。我会为你介绍一位很好的心理咨询师，由她给你做。"

朋友说："这个病人不是我，是我一位同事的亲戚的朋友的孩子。说实话，我并不认识这个病人，和我也没有多么密切的关系，人家信任我，我才来穿针引线。"

我说："你真是古道热肠，拐了这么多的弯，还把你急成这样。给你个小小的纠正，来做心理咨询的人不是病人，我们通常称他们为来访者。"

朋友说："这有什么很大的不同吗？叫病人比较顺嘴。"

我说："很多人来做心理咨询，并不是因为有了心理疾病，而是为了寻求更好的发展潜能和更亲密的人际关系。"

The Magic Word

An old friend called one day, complaining that she could not get hold of me, that I had disappeared. I teased, asking if this was an invitation to dinner, assuring her that I was available. She said this is no joking matter, that when the problem is taken care of, she would give me a banquet, but that right now, we could not afford to lose a moment. I noticed the urgency in her voice, and asked her to get to the point.

She sighed and said that she needed my services for psychological counseling. I said I am glad that nowadays people are more concerned over their psychological health but added that I could not deal with her case as we were friends, and that I would recommend her to a very good counselor. She replied No, she was not calling on her own account. She told me that this "patient" was the son of the friend of a relative of her co-worker, and that she herself was in the dark about the boy's problem, but had offered to help locate a good counselor.

I admired her generosity in offering help but pointed out that people who come to us are not "patients," that we usually refer to them as visitors.

My friend said, "what's the difference? Patient is easier to pronounce."

I told her that many people come for psychological counseling, not because of illness, but looking for ways to draw out their full potential, or learning how to strengthen personal relationships.

朋友说："但我说的这个孩子确确实实是病了。当然不是身体上的病，他的身体棒得能参加奥运会，但却不肯去上学。再有两个月就要高考了，这是多么关键的时刻，可他说不上就不上了，谁劝也没用。一家人急得爸爸要跳楼妈妈要上吊，他却无动于衷，整天把自己关在屋里玩电脑，任谁都不见。家里人急着要找心理医生，但这个孩子主意太大了，根本就不答应去。后来，他家里人找到我，让我跟你联系。那孩子说如果是毕淑敏亲自接待他，他就前来咨询。现在总算联系上了，你万不能推托。你什么时候有时间呢？让他父母带着他来见你……"

我一边听着朋友的述说，一边查看工作日程表。最近的每一个时段都安排得满满的，只有7天后的傍晚有一小时的空闲。

我把这个时间段告知了朋友，请她问问那位中学生届时有没有空。

朋友大包大揽道："只要你能抽出时间，那边还有什么好说的？他们一定会来的。"

我很严肃地对她说："请你一定把我的原话传过去。第一，要再次确认那位中学生是自己愿意来谈谈他的想法，而不是被父母强迫而来的。第二，征询那个时间对他合不合适。如果他有重要的事情，我们还可以再约另外的时间。第三句话就不必传了，只和你有关。"

朋友说："前两件我都会原汁原味地传达到。只是这第三句话是什么，我很想知道。怎么把我这个穿针引线的人也包括进去了？"

我说："第三句话就是，你的任务就到此为止了。因为这种特殊的就诊方式，你已经卷入了开头部分。关于进展和结尾，恕我保密。你若是好奇或是其他原因追问我下文，我会拒绝回答。到时候，请你不要生气。不是我不理睬你，友情归友情，工作是工作，保密是原则问题，祈请见谅。"

My friend insisted that in this case, the person under consideration was sick indeed, not physically—physically he could qualify for the Olympics—but just that he refused to go to school. "Imagine," she said, "just two months away from the National Examination for Higher Education. His whole future is hanging on it. And he is playing hooky. The family is in an uproar. His father is hopping mad, and could throw himself out the window. His mother wants to hang herself. The boy is totally unconcerned, and locks himself in his room, playing computer games. Nothing will move him. The family found a counseling psychologist, but the boy refuses to show up. Now the family has found me—they are friendly with the relative of one of my co-workers, and I am asked to contact you. The boy had said that he doesn't mind talking to Bi Shumin in person. Now you have the case before you and you are not going to get out of it. Give me a date and his parents will bring him over!"

I glanced at my schedule as my friend was talking on the phone and it seemed that there was a slot for an hour, seven days away. I asked her to make an appointment for the boy. My friend was elated.

I emphasized two points.

One, that the visit should be voluntary on the boy's part. Two, that we could always change the date to fit his schedule.

Additionally, a third point was that "From now on," I told my friend, "your job is done; you need not involve yourself any longer. Our friendship is there, but my work involves another person's privacy, and I cannot bring it into our conversation. I hope you will understand."

朋友说："好，我把你的话传到就算使命截止。我会尊重你们的工作规定。"

一周后的傍晚，一对衣着光鲜的夫妻押着儿子来了。我之所以用了"押"这个词，是因为夫妇俩一左一右贴身护卫着那个高大的年轻人，好像怕犯人逃跑的衙役。年轻人走进咨询室的时候，他们俩也想一并挤入。

接待人员递给我咨询表格，轻声对他们说："你们并不是整个家庭接受咨询。"

年轻人说："对，这是我一个人的事。"说完，他懒懒散散走进了咨询室，一屁股坐在沙发上，目光直率地打量着我。我也打量着他。

他叫阿伦，身高大约一米八三，双脚不是像旁人那样安稳地倚着沙发腿放置，而是笔直地伸出去，运动鞋像两只肮脏的小船翘在地板中央。他身上和头发里发出浓烈的龌龊汗气，让人疑心置身于一家小饭馆的烂鸡毛和果皮堆的混合物旁。

我抑制住反胃的感觉，不动声色地等着他。

"你为什么不先说话？"他很有几分挑衅地开始了。

我说："为什么我要先说话呢？这里是心理咨询室，是你来找的我，当然需要你先说出理由了。"

他突然就笑了，露出很整齐却一点也不白的牙齿，说："你说得也有几分道理啊。不过，是他们要我来见你的。"

我问："他们是谁？"

阿伦歪了歪鼻子，用鼻尖点向候诊室的方向，在墙的那一边，走动着他焦灼不安的父母。

我表示明白他的所指，把话题荡开，问道："你好像比他们的个子都要高？"

My friend said she understood, that she will pass on my message to the parents, and that they will respect the rules of the game.

A week later the parents, formally dressed for the occasion, showed up in my office, with their son in tow. Mom and Dad stood on either side of the boy, as if holding him under guard. My staff handed me the question-and-answer sheet, and turned to the parents, telling them that this was not a family consultation. The young man, squeezed between his parents, stated: "Right, this concerns me only." Saying which, he shook them off and walked casually into the consulting room. Needing no invitation, he plumped down on the sofa, eyes fixed on me. I also had my eyes on him.

Alan was about 1.83 meters tall. His long legs were stretched straight out, while his shoes were like two dirty little boats stranded in the middle of the floor. His clothes and his hair stank like the trash heaps outside those sleazy little street eateries. I controlled my repugnance and waited for him to break the ice.

"Why don't you speak first," he said, tantalizingly.

I retorted, "Why should I speak first? This is a room for counseling. You looked me up. Now give me a reason."

He burst out laughing, showing irregular white teeth. "*They* wanted me to come," he said.

"Who are they?"

He turned, pointing his nose in the direction of the reception room outside, where his parents were probably on pins and needles, holding their breath.

I understood his body language and changed the subject, saying: "You are much taller than your parents."

他好像受到了莫大的夸奖，说："是啊，我比他们都高。"

我说："力气好像也要比他们大啊！"

阿伦很肯定地点头说："那是当然啦！我在三年前掰腕子就可以胜过我父亲了。"

我把话题一转："如果你不愿意来，你的父母是无法强迫你到心理咨询师这里来的。"

阿伦愣了一下，说："对，我是自愿的。"

我说："既然你是自愿来的，那你有什么问题要讨论呢？"

阿伦说："我其实没有问题，是他们觉得我有问题。我不过是上上网，玩玩电子游戏，有什么了不起的？"

我不想跟阿伦在到底是谁有问题的问题上争执不休。因为第一次咨询的任务，最主要是咨询师要和来访者建立起良好的关系，培养起信任感并了解情况。我说："你一天上网的时间是多少呢？"

他说："大约18个小时吧。"

我无法掩饰自己的惊讶，问道："那你何时吃饭，何时睡觉呢？"

阿伦说："饿了就吃，一顿饭大约用3分钟。实在熬不住了就睡，每次睡15分钟再起来战斗。我发现人一天睡5个小时就足够了，说睡8小时那是农耕时代的懒惰。"

我说："首先恭喜你……"

我的话还没有说完，就被阿伦打断了："您不是在说反话吧？"

我很惊奇地反问他："你从哪里觉得我是在说反话呢？"

阿伦说："所有的人知道我这样的作息时间之后，都说我鬼迷心窍，哪能一天只睡5小时呢？"

He took it as a compliment: "Yeah, I am taller than both of them,"

"And stronger too, I presume,"

Alan nodded vigorously, "Of course! I had beaten my dad at wrist-wrestling three years ago!"

I changed the subject saying: "So if you did not want any counseling, they can't make you..."

Alan seemed startled, then said: "True, I came here of my own free will."

"Since you wanted to see me, why don't you come to the point?"

"Actually, I do not have any problems. *They* think that I have a problem. All I did was to get on the internet, play a few computer games. And so what if I did?"

I did not want to get mired in the problem of who has a problem. In a first consultation, the most important thing is to build trust and pinpoint the issue.

I asked: "How many hours daily do you spend on the internet?"

"About eighteen hours," he said.

I could not hide my surprise. "Then when do you eat? And when do you sleep?"

He said, "I eat when I am hungry; it takes about three minutes. When I am dead tired, I take a nap. A fifteen-minute nap would set me up for a new round of games. I find that five hours of sleep is more than enough to set you up for the day. The eight-hour sleep is a relic of the rural way of life."

I said: "First of all, I must compliment you..."

Before I could finish the sentence, Alan broke in saying: "You are pulling my leg."

"What makes you think so?"

"Well, people who know my lifestyle usually say that I am off my rocker, that no one survives on five hours of sleep a day."

我说："我要恭喜你的也正是这一点。因为通常的人是需要每天睡眠8小时，如果你进行了正常的工作学习而只需要5小时睡眠就能恢复精力，这当然是值得庆贺的事情。每天能节约出3小时，一辈子就能节约出若干岁月，你要比别人富余很多时间呢，当然可喜可贺。"

阿伦点点头，看来相信我说的是真心话。

我紧接着问道："那你何时上学做功课呢？"

阿伦皱起眉头说："您是真不知道还是假装不知道呢？我已经整整28天不去上学了。"

我发现当他说到"28天"这个日子的时候，眼睫毛低垂了下去。我说："看来，你还是非常在意上学这件事的。"

他立刻抗议道："谁说的？我再也不想回到学校了，那是我的伤心之地。"

我说："你连每一天都计算得这样清楚，当然是重视了。只是我不知道，在28天以前发生了什么重大的事情，让你做出了不再上学的决定，直到今天还这样愤怒伤感？"

阿伦很警觉地说："你到学校调查过我了？"

这回轮到我笑起来说："你真是高估了我。你以为我是克格勃？我哪有那个本事！"

阿伦还是放不下他的戒心，说："那你怎么知道28天以前发生过什么重大的事情？"

"Well, that is precisely why you deserve a compliment. Most people survive on eight hours of sleep per day. But you survive on five and still lead a normal life. So you are saving up many work days in your lifetime, and isn't that something to be complimented on?"

Alan nodded, evidently taking my point.

"Now when do you do your homework?"

Alan frowned: "Are you really in the dark, or are you just kidding? Don't you know that I have not shown up at school for the last twenty-eight days?"

I noticed that when he was talking, his voice dropped solemnly at the mention of the word *school*.

"Obviously," I said, "To you, school is something that you take very seriously."

Alan protested, "No, not at all. I am not going back to school. I might as well say: school is my *Heartbreak House*."

"But you have reckoned the days so carefully; it means that you care. My only problem is this—what exactly happened that hurt you so badly as to keep you away from school for the last twenty-eight days? And you are still hurting."

Alan asked cautiously: "You went to my school and did some probing?"

I laughed: "What do you take me for? The KGB?"

Alan was still on the defensive: "Then how do you know that something momentous had happened twenty-eight days ago?"

我收起笑容，说："能让你这么一个身高体壮、智力发达、反应灵敏的年轻人做出不上学的决定，当然是一件重大的事情啦！"

阿伦说："你猜得不错。28天之前，正好是我们模拟报高考志愿的时候。我看到发下来的报名表，想也没想就填上了'清华大学'。当然了，我的成绩距离上清华还有很大的差距，但我想，距离考试还有几个月的时间，谁说我就不能创造出点奇迹呢？再有，气可鼓而不可泄呢，这也是兵法中常常教导我们的策略嘛！"

"没想到代课老师走到我面前，斜眼看了看我的志愿说，就你这德行也想报清华，你以为清华是自由市场啊？

"那天正好我们的班主任因病没来，要是班主任在，也许就不会出事了。这位代课老师因为我有一次打篮球没看见她，忘了问好，就被她记了仇。

"我说怎么啦，清华就不能报了？

"老师说，也不看看自己的成色，别给学校丢人了，这样的报考单送到区里做摸底统计，人家不说你不知天高地厚，反倒说是老师没教会你量力而行……

"如果老师单单说到这里就停止，我也就忍气吞声了。学校里，老师挖苦学生是天经地义的事，我们都麻木了，我低下了头。老师不依不饶，她撇着嘴说，就凭你这样的人还想为校争光，那我就大头朝下横着走！"

I put away my smile and said in all seriousness: "Something momentous must have happened to make a tall, smart, sensitive guy like you decide to quit school."

Alan conceded, "You have guessed right. Twenty-eight days ago, we the graduating class had to fill a trial form for application to college. I picked up the sheet of paper and casually filled in Tsinghua University. To judge by my grades at the time, obviously I could not expect to get into Tsinghua. But there were still a couple of months ahead. If I work hard, who is to say that a miracle is absolutely out of reach? Our aspirations should be encouraged, right? Not dashed! Isn't it one of the rules of Sunzi's ancient *Art of War*? But you know what? The teacher strolled over and took a glance at my application form.'Who do you think you are,' she scolded, 'applying for Tsinghua University? Do you take Tsinghua for a flea market?'

"What followed would never have happened if our regular teacher had been there. But he had called in sick and left this stand-in to take over." Alan confided, "Once when I was playing basket-ball, she passed by and I didn't greet her. She had never forgiven me for the slight."

Alan continued his narrative. "To her taunt, I had replied: 'And so what? Why can't I apply to Tsinghua if I want to?' And she had shot back: 'Don't you have a sense of reality? You will be a laughing stock. And what's more, the District Education Office will blame us for not advising you kids to make sensible choices based on self-knowledge!'

"If the stand-in teacher had stopped there," Alan paused, "nothing would have happened. Over time we students have been numbed by the scathing remarks of the teachers, and on that occasion I just held my tongue. But she could not stop. Twisting her mouth in derision, she went on to say: 'If the likes of you could bring credit to our school, I will get on my hands and knees and crawl sideways!'"

听到这里，我忍不住插话道："这位老师如此伤害你的自尊心，我听了很生气。"

阿伦没理我，自顾自说下去。

"不知为什么，老师这句话强烈地刺激了我，我一想起面目可憎的老师能像个螃蟹似的，头抵着土在地上爬行，就不由自主地哈哈大笑。老师摸不着头脑，但是能感觉到我的笑声和她有关，就厉声命令我不要笑。但我依旧大笑不止，她束手无策。那天我笑得天昏地暗，从学校一直笑回了家，闹得父母很吃惊，以为我考了100分。

"我走火入魔似的陷入了这种想象之中，但是要让老师真的趴在地上，是有条件的，我必得为校争光。真的考上清华吗？我没有这个把握，若是考不上，岂不验证了老师对我的评判？我就滋生了放弃高考的念头。

"一场考试，如果我根本就没有参加，就像武林高手不曾刀光剑影华山论剑，你就无法说谁是武林第一。但是放弃了高考，我用什么来证明自己呢？我想到了网络游戏……"

说到这里，阿伦抬起头，问道："您玩网络游戏吗？"

我老老实实地回答："不玩。我老眼昏花的，根本就反应不过来。"

阿伦同情加惋惜地叹口气，说："那您也一定不知道'魔兽''部落''联盟'这些术语了？"

我说："真的很遗憾，我不知道。但我很想向你学习。"

我说的是真心话。既然我的来访者是这方面的高手，既然他沉迷于网络不能自拔，我当然要向他请教，我要走入他的世界，我要感同身受地体验到他的快乐和迷惘，我必须了解到第一手的资料和感受。

I could not help exclaiming, "How could the teacher be so rude?"

Alan ignored my remark and went on: "The stand-in's words struck a chord. I have an image of her, head down, moving sideways like a giant crab, and I burst out laughing. She was confused by my strange behavior. She sensed that it had something to do with her, something decidedly not complimentary, and she ordered me to stop. But I was convulsed with laughter and could not stop. I laughed all the way home. My parents thought that I must have gotten an A being so happy."

Alan continued: "I was fixated by the image of the teacher crawling sideways like a crab. But there was a string attached—she would crawl like a crab only if *the likes of me* could bring credit to the school. But how can I bring credit to the school? Certainly not by getting into Tsinghua University! It occurred to me that if I refuse to take the test for higher education, then the issue of failure would not come up, and the teacher would not be vindicated. It would be like the case of those martial art heroes of the old days—if they do not show up for a contest, they cannot be proven to have lost! And then a new question arose—if I give up the National Test for Higher Education, what else can I do to prove myself and bring credit to our school? Answer—computer games!"

Here, Alan stopped and asked me, "Do you play computer games online?"

"Oh no! Not at my age! My eyesight is bad, I can't keep up…"

Alan sighed in sympathy. He asked me: "So you don't know those terms such as World of Warcraft or The Alliance?" I confessed that I had no idea what he was talking about, but added that I am willing to learn if he teaches me.

I really meant what I said. Since my visitor excels at computer games and is totally immersed, I certainly should learn something from him. I must enter his private space and share his joy and his confusion. I must gather data. In a word, I must get under his skin for first-hand knowledge of what he was feeling.

阿伦说："那我就要向您进行一番普及教育了。"他说着，有点儿似信非信地看着我。

我马上双手抱拳，很恭敬地说："阿老师，请你收下我这个学生。只是我年纪大了，脑袋瓜也不大好使，还请老师耐心细致地讲解，不要嫌弃我笨。如果有不明白的地方，我会提出来，也请老师深入浅出地回答。"

他快活地笑起来，说："我一定会耐心传授的。"说完，他就一本正经地向我解释起经典游戏的玩法。我非常认真地听他讲授，重要的地方还做笔记。说实话，专心致志的劲头，只有当年在医学院做学生听教授讲课的时候才有这般毕恭毕敬。

交流平稳地推进着，离结束只有10分钟时间了。按照咨询的惯例，我要进入"包扎"阶段。也许在不同的流派里，对于这段时间的掌握和命名各有不同，但我还是很喜欢用"包扎"这个术语。

咨询的过程，在某种程度上就是打开了来访者的创伤，在来访者离去之前，一个负责任的心理咨询师要把这伤口消毒与缝合，让来访者在走出咨询室的时候不再流血和呻吟。心理创伤和生理创伤一样，陈年旧疾和深入的刀口，都不是一朝一夕可以愈合如初的。

心理咨询师要有足够的耐性和准备，第一次咨询主要是建立起真诚的信任关系和了解情况，其余的工作来日方长。

我说："谢谢你如此精彩的讲解，现在，我对网络游戏有了一些了解。"

阿伦轻快地笑起来，说："能和您这样谈话，真是很愉快啊。我还要再告诉您一个重要的秘密，我就要代表中国和韩国的选手比赛，如果我们赢了，那就真是为国争光了！"

我伸出手来祝贺他说："你在游戏中充满了爱国精神。"

他紧紧地握住了我的手，说："您说的是真心话吗？"

Alan said, "Well, we have to start with some basic info," saying which, he looked at me, unsure of my reaction.

I put my two hands together in the traditional pose of humility, saying: "Teacher Alan, please accept me as your student. I am advanced in years and rather slow. But I will ask whenever I have a problem, and I hope you the teacher will help me understand."

Alan laughed happily, saying: "I promise to be patient!" Saying which, he started to explain to me in earnest the rules of the popular computer games, and I took notes. To be honest, I have never been so respectfully studious since my student days in medical school.

This private instruction proceeded until it was ten minutes to the end of his consulting session. According to protocol, this was the time for "bandages." Different schools of psychological counseling have different terms to denote this final wrapping up. I like the term "bandages." The process of asking questions is like probing into the wound of the visitor. By the end of the session, the counselor should clean up the wound and bind it up. One must not let the visitor leave the consulting room still bleeding. The psychological damage, however, like a physical wound, does not heal at a snap of the fingers. The counselor must be prepared for the long haul. The first session is usually for building trust and collecting basic information. The rest is left to time.

I thanked Alan, saying that I have grasped the rudiments of computer games.

Alan laughed happily, saying that it was a pleasure to talk to me. He then confided: "I am representing China for a computer game contest against Korea team. If I win, it will be winning honor for my country!"

I shook his hand and said, "You will amply show your patriotism just by participating in the contest!"

Alan held my hand, saying: "Do you really mean what you are saying?"

我说：“当然，你可以使劲握住我的手，你可以感觉到我的手的力量。如果我的话是假的，我会退缩。”

阿伦真的握住了我的手，我感觉到他的手在轻轻地发抖。

分手的时间到了，我对阿伦说：“谢谢你对我的信任，告知我那么多的知心话，我会为你保守秘密的。也谢谢你耐心地为我这样一个游戏盲讲解游戏，让我对此有了一定的了解。我希望在下个星期的这个时间能够看到你来，咱们还要讨论为国争光的问题呢！”

阿伦脸上的神色突然变得让人捉摸不透。他对我说：“原谅我下个星期的这个时间不能来到您这里了。”

我尊重阿伦的意见，因为如果来访者自己不愿意咨询了，无论咨询师多么有信心也无法继续施行帮助计划了。

我表示理解地点点头。

阿伦突然扬起了眉毛，说：“下个星期的这个时候，我想我是在学校上晚自习吧。您知道，毕业班的功课是非常吃紧的。”

我大吃一惊。说实话，在整个咨询过程里，我们不曾探讨上课的事，我认为时机未到。

阿伦是个无比聪明的孩子，他看出了我的困惑，说：“我知道爸爸妈妈领我来的意思，谢谢您没有说过一句让我回去上课的话。在来的路上我就想好了，如果您也千篇一律地劝我的话，我会扭头就走。谢谢您，什么也没说。您向我讨教游戏的玩法，我很感动。

“从小到大，还没有一个成年人如此虚心地向我求教过，这样耐心地听我说话。还有，您最后祝愿我为国争光，我非常高兴，您终于理解我的不上学其实只是想证明自己是有能力做一些事情并且能做好的。

I said: "Squeeze my hand as tightly as you can. If I didn't mean what I said, I will not be able to hold out."

Alan held my hand, and I felt a slight tremor.

The session was up. It was time to say goodbye.

I thanked Alan for trusting me and telling me his innermost thoughts, and I promised to keep them to myself. I also thanked him for introducing me to computer games, and ended by saying: "I hope to see you at the same time next week and we might discuss the issue of 'winning honor for the country.'"

The expression of Alan's face was inscrutable. He murmured: "This hour next week? I am afraid I can't make it."

I nodded, respecting Alan's decision. Nothing will work if the advisee is unwilling to be advised. Alan lifted his eyebrows, saying: "This time next week, I will be in school—the evening self-study session. You know how tense it is for the graduating class..."

I was happily surprised. To tell you the truth, all through the session, I had never mentioned school. I had thought that the time was not yet ripe.

Alan is a smart kid and he saw through my confusion. "I know why my parents brought me here. Thank you for not saying a word about going back to school. My mind was made up on my way here. If you had repeated the rigmarole about going back to school, I would have turned on my heels and left. I am thankful to you for not mentioning a word about school. You asked me to teach you computer games and I was very flattered.

"All my life, no adult had ever asked me for help so modestly, or had listened to me so patiently. What's more, you wished me success in winning glory for our country. It made me so happy. You understood that I kept away from school in order to prove that I was capable of doing something significant on my own.

“对了，您还表示了对那个老师的愤慨，让我觉得很开心，觉得自己不再孤独和愚蠢……现在，我不需要再用网络游戏来证明什么给那个老师看了，我要回到书本中去了。我知道这也是您希望的，只是您没有说出来罢了。”

我们紧紧握手，这一次，他的手掌都是汗水，但不再抖动。

过了暑假，那位朋友跟我说：“你用了什么法子让那个网络成瘾的孩子改邪归正的？他的父母非常感谢你，因为他考上了重点大学，真是考出了最好的成绩呢！他们想请你吃饭，邀我作陪。”

我说：“咱们可是有言在先的，我不能向你透露任何相关的信息，也不能赴宴。如果你馋虫作怪，我来请你吃饭好了。”

朋友说：“我看他们感谢你还不是最主要的目的，主要是想探听出你究竟跟他们的儿子说了点什么，能有这么大的功效。”

我说：“那一天，我说得很少，阿伦说得很多。其余的，无可奉告。”

"And you had expressed disapproval of that stand-in teacher. I felt vindicated. I stopped feeling stupid and isolated. Now I do not need to prove myself to that stand-in. What I need now is to get back to school. I know it is what you want for me."

We had a good handshake.

At the end of that summer, my friend called to ask what I had done to make the boy mend his ways. She said that the parents are grateful and want to take me out for dinner. She added that the boy had taken the National Test and had gotten into a good college.

I reminded my friend that we had had a prior agreement that I do not reveal the contents of private conversation, and that I do not accept personal invitations. I added: "If you want to do lunch, I will treat."

Finally my friend confessed, "I suspect that taking you out was a tagged-on item—what the parents really wanted to know was how you managed to turn him around..."

"All I can tell you," I replied, "is that during the session, I barely opened my mouth. It was Alan who did the talking."

出卖冥位的女生

来访者是一位中年女子，名叫鞠鸣凤。衣着得体，在她的登记表“心理咨询事由”一栏中，填写的是“人为什么要出卖冥位”？结尾处的问号又长又大，像一根生了锈的铁锚直击海底。

我看着这问号愣了一会儿。别说她不知道这个问题的答案，我连冥位是什么东西都不清楚。好在，我并不着急。世界上的万物就是如此复杂，一个咨询师不可能什么都知道。这不是咨询师的耻辱，只是一个真实。不过，世界上的万物又都是有规律可循的，只要跟随着来访者的脚步，我们就有可能一同到达彼岸。

鞠鸣凤坐下后，第一句话是：“您知道什么是冥位吗？”

我老老实实地回答：“不知道。很希望您告诉我。”

鞠鸣凤说：“冥位就是埋葬死人的地方。可以是一块地，也可以是一棵树，一个花坛，也可能是灵塔上的一个格子。”

我明白了一点点，但也更糊涂了。我说：“难道一个人可以埋在这么多的地方吗？”

鞠鸣凤说：“不是。也许是我没说清楚。每个人死后只占据一个冥位，冥位是商品。要知道冥位是可以买卖的。现在房地产涨价，阴间的地盘也紧张起来，所以，有些人成了殡葬业的推销员，就是出卖冥位的人。”

The Girl who Sold Burial Plots

The new caller was a middle-aged woman, tastefully dressed, named Ju Mingfeng.

When filling the form for the subject to be consulted on, she had written: "Why do people sell burial plots?" The question was followed by a huge question mark, with the dot looking like a huge rusted old anchor at the bottom of the deep blue sea.

I stared at the writing. Not only did I have no answer to the question—actually I had no clue as to what she was talking about, but I didn't let it bother me—one should not expect a counseling psychologist to have an answer for everything. My ignorance is nothing to be ashamed of. It is just a fact that one has to accept. All things under heaven have their own rules for self-revelation, after all. So if we follow in the footsteps of the visitor, we will eventually get to the bottom of the matter.

Ju Mingfeng sat down and shot a question at me: "Do you know what is meant by burial plots?"

"No, I don't. Supposing you tell me?"

"Burial plots are where the dead are buried," she said, "It could be a plot of land, or a tree, or a flower bed, or a slot up in a funerary tower."

I got the point but was still confused by the multitude of choices—could one be buried in so many different places?

"Let me clarify," she hastened to add, "People have only one choice, of course. Burial plots are items for sale, like any other commodity. Now that real estate prices have hit the roof, the market for burial plots is also hot. This explains the appearance of sales representatives, who would push burial plots on people, like any other commodity."

原来是这样。大千世界，真是无奇不有啊。我说："谢谢您告诉我这样的知识。原来出卖冥位是世上的新行当。"

鞠鸣凤说："本来这行当新呀旧的跟我没关系，可没想到我的女儿鞠小凤卷了进去。一天像着了魔似的推销冥位……"

我有点儿吃惊。鞠女士的年纪也就40出头，她的女儿能有多大呢？不到20岁吧？小小年纪就成了推销埋葬死人骨灰的业务员，这太匪夷所思了吧？鞠鸣凤看出了我的疑惑，说："是啊，她还在上高中。我今天来找您，就是为了解决她的问题。现在，我马上出去，把她换进来，让她自己跟您说说到底是怎么一回事吧！"说完，起身走出门去。外面负责接待的工作人员不知发生了何事，以为她对我的咨询不满，半路上扬长而去。

我轻轻摆摆手，示意工作人员不要阻拦。

这真是我工作经验中的一个新鲜事。咨询过程居然像篮球比赛，玩起了半路换人。我且要看看这个正上高中，却成了冥位推销员的小姑娘，是个怎样奇特的人。或许穿着哈韩哈日的肥腿裤吧？或者衣衫褴褛头发被发胶粘成图钉状？或者一身迷彩戴着贝雷帽手握仿真枪……

我所有的想象都在现实的面前碰得粉碎。鞠小凤身材高挑，健康活泼，身穿一套天蓝色夹有雪白条纹的校服，一步三跳地走了进来。她毫不认生地一屁股坐在她妈妈刚才的位置上，说："嗨！听我妈妈一讲，您一定以为我是个怪物吧？其实，我非常正常。本来不打算到您这儿来的，后来一想，我也没见过心理咨询师是什么样的，开拓一下自己的见识也很重要。再说，没准儿我还能向您推销一个两个冥位呢！"

我一下子目瞪口呆，没想到我居然成了她的推销对象。

So that was it! Nowadays, one should not be surprised at anything new. I said: "Thank you for the information. So selling burial plots has become a new line of business?"

Mrs. Ju said: "Really, the business, whatever it is, has nothing to do with me—until my daughter got involved. She has thrown herself into selling burial plots like a girl possessed."

I was surprised. Mrs. Ju looked about forty. Her daughter could not be over twenty. How could she throw herself into selling burial plots?

Mrs. Ju saw my incredulity. "It is true," she said, "she is still in junior high. That is precisely why I am here. I will let her tell you everything in her own words." So saying, Mrs. Ju stalked out of the room. The staff was surprised to see her leave, wondering if the lady had been offended in any way. I signaled my staff not to stop her exit.

This is a totally new experience in my professional life as a counseling psychologist—as if one could change players half-way as in a basketball match!

I looked forward to meeting this phenomenon of a high school girl selling burial plots, curious as to the kind of girl she could be. Would she appear in wide pants in the Japanese/Korean style? Or dressed in rags with her hair pinned to her head? Or in battle fatigues and beret, sporting a shotgun?

All my conjectures were shattered by the real person.

Ju Xiaofeng bounced into the room, tall and healthy, the typical school girl, in blue-and-white uniform. She sat down in the seat that her mom had just left and announced: "By my mom's account, you must expect me to be a monster. Actually I am just like everybody else. I didn't want to come. Then I thought to myself, I have never had psychological counseling. Why not give it a try! It would widen my horizon. Besides, I might be able to sell a plot or two."

I was dumbstruck, never imagining that I would become the target of a sales pitch.

我调整了一下思绪，说：“小凤，谢谢你。我还真没想到要为自己置办一处冥位的问题。”

鞠小凤丝毫不受打击，依旧兴致勃勃地说：“没想到不要紧，现在开始想想也来得及。您知道，伟大领袖毛主席说过，人必有一死。死了以后，您住在哪里呢？总要有一个地方吧？要么变成一棵树，要么变成一朵花，要么就安安静静地睡在泥土里……您现在就可以选择。对了。老师，我现在就向您介绍一个好地方，山清水秀的，空气可好了。最主要的是邻居好……”

“邻居好？”我不由得失声追问。

“对啊！”鞠小凤兴头正高，眉飞色舞地说，“您以为灵魂就不需要邻居了吗？一样需要，甚至更重要。因为灵魂像风一样，经常到外面去飞翔，自己的家就要托邻居照料。这处冥位，旁边都是知识分子，有大学教授啊，有律师和医生啊，最有意思的是，还有一位是大使，这样您还可以听到很多外国的故事……”

鞠小凤说得津津有味，我跟着她的语调，真的想到了一片开阔的青草地，鸟语花香，然后仿佛看到一群西服革履的人，正谈笑风生。

天啊，这个小姑娘真是不简单，连我这把年纪的人，都被她蛊惑。

“怎么样？买一个冥位吧？”鞠小凤问我。

我赶紧回到自己的工作状态，对她说：“你干这行多长时间了？”

鞠小凤说：“没多久。我是偶然知道这个消息的。其实并不复杂，都是正规陵园，手续齐全。我们推销出一套冥位，就能有一定的提成。我也不会耽误学习。”

I collected my thoughts and said, "Thank you. So far I have no intention of buying a burial plot."

Nothing daunted, Ju Xiaofeng went on with great gusto: "It is never too late; you can start thinking about it now. Remember, Chairman Mao had said: 'Man must die...' Since all men must die, we should all consider where to go after death. One must have a place to stay. You could be a tree, you could be a flower; or you would lie in peace underground. Whatever it is, you need space. By the way, I can recommend something right now. An ideal spot: green hills, sparkling brook, pure air... What's more, good neighbors, too!"

"Neighbors!" I could not help exclaiming.

"Exactly! Are you assuming that you don't need neighbors anymore once you are dead? Wrong! You need them even more! And this is because your soul is light as the wind, flitting to and fro. And when you are out, you need neighbors to keep an eye. Fortunately the spot I have in mind for you is chock full of members of the educated elite—doctors, professors, lawyers and so on, even an ambassador! You will be able to hear a lot of foreign scandals."

Ju Xiaofeng was totally immersed in her sales pitch, and I was led into a scenario of a group of people in shirt and tie, carrying on an animated conversation against a background of wide open spaces with green grass, colorful flowers and chirping birds.

Goodness, this little girl is really something! She almost talked me into the mood.

"Now, what do you think? Come on, buy a plot," she urged.

I went back into my role as a counselor and asked her: "How long have you been doing this?"

"Not long," she said, "I stumbled into it by accident. Don't think of it as something messy; it is quite straightforward. There is this public cemetery, all papers in order. Every time I sell a plot, I get a kickback. It doesn't interfere with my schoolwork at all."

我说："你做这个工作，是为了挣钱吗？"

鞠小凤说："挣钱肯定是一个原因。像我们这个年纪的女孩子，都是向家里要钱的，我第一次拿到我的提成时，非常高兴。因为这证明我的能力。但是，钱并不是最重要的。"

我点点头表示理解，追问："那么，什么是最重要的呢？"

鞠小凤好像很不愿意触及这个问题，说："一定是我妈妈跟你说了很多坏话。好像我一个女孩子干这个事，是大逆不道。她非常害怕死亡，还说我以后长大了，要是让人知道我曾经干过这个行当，肯定会嫁不出去了。可是，我不怕。我不害怕死亡。"

鞠小凤说这些话的时候，神色迷离，目光弥散，一下子失魂落魄。

按说一个女孩子不害怕死亡，是难得的勇敢，可我总觉得有什么地方不对头。不过从这个方向探寻她的内在世界，难以进入。我略一沉思，发现了一个问题——她妈妈叫鞠鸣凤，她叫鞠小凤。按说鞠这个姓氏并不常见，难道说一家三口人都姓鞠不成吗？如果不是这样，鞠小凤就是从母姓，那么鞠小凤的父亲到哪里去了呢？

我决定从这个方向入手。我说："小凤，我看你对死亡的认识很豁达，如果你不介意的话。能同我谈谈你的父亲吗？"

鞠小凤说："我妈妈没有跟您说吗？"

我说："没有。她就只是说到了你。"

"Are you doing this because you need the money?"

"Money is part of it, of course," she owned, "Girls like me always rely on family for pocket money. The first time I was paid, I was so happy. I had proven myself! But still, money is not top priority."

I nodded in understanding, and pursued: "If not money, what is at stake?"

Ju Xiaofeng did not wish to go into the problem. She said: "My mother must have been bad-mouthing me, as if I am doing something shameful. She is living in daily fear of death. Besides, she is afraid that no one will marry me if I am known to have done this. But I don't care. And I am not afraid of death, either..."

As Ju Xiaofeng was saying the above, however, her high spirits seemed dashed, her eyes wandered vaguely, and she suddenly lost her verve.

Girls her age must be armed with rare courage not to fear death. But I still felt something missing. I felt in my bones that I will get nowhere following this line of conversation. It will never lead into her inner world. Then something struck me as strange. Her mother is Ju Mingfeng, and she is called Ju Xiaofeng. Among the list of Chinese family names, the ideogram Ju is unusual and rarely appears. Is it possible that the girl's father is also surnamed Ju, so that she is a Ju after her father, while her mom's last name also happens to be Ju? It is unlikely. The only explanation is that she is taking her mother's last name. Then what about the father?

I decided to shift to a new angle. "Xiao Ju," I said, "you really hold a very broad-minded view towards death. Would you like to tell me something about your father?"

"Did my mother not say anything about my father?"

"No, we had only talked about you."

鞠小凤平静地说："我的亲生父亲在我很小的时候，就在一次飞机失事中去世了。当时飞机一头扎到海里，所有的人尸骨无存。后来，我妈妈就带着我改嫁了，继父对我很好。嗯，很简单，就是这样。我妈妈又把我的姓改成了她的姓。从此，我的亲生父亲在这个世界上，就没有任何痕迹了。"

我发觉鞠小凤把"尸骨无存""任何痕迹"几个字咬得很重。如果把她这段话比作一块木板，那么，这几个词，就像木板上凸起的木疤，显而易见，触目惊心。

我基本上找到了症结。我说："你非常思念你的父亲？"

鞠小凤的眼眶一下子红了，说："无论我的继父对我多好，可是，我的骨头我的牙齿我的头发，不是他给我的，是那个在这个世界上消失得无影无踪的人给我的。我非常想念他。可是，我不敢让我妈妈发现，那样，她就会觉得委屈了我。其实，那不是她的过错。我只是用我的方式，纪念我父亲。"

我紧紧跟上一句："什么叫作你的方式？"

鞠小凤说："那就是思索和死亡有关的一切。比如，我认为死后是有灵魂的，我认为人是应该留下一点儿痕迹的。不然的话，我们的哀伤就找不到地方寄托。"

我知道，我们已经渐渐逼近了问题的核心。

我说："你觉得哪些可以称之为痕迹呢？"

鞠小凤说："比如一块土地，比如一朵花，比如一棵树，不能什么都没有。那样，活着的人会受不了的。"

Ju Xiaofeng told me quietly: "My father died in an airplane accident when I was a tiny baby. The plane dived into the sea, taking with it all passengers and crew. My mother remarried later, and my stepfather has always been very good to me. That's it. Really simple. Some time later, my mother made me take her last name, Ju. Thus my own father is completely wiped out of existence, leaving not a single trace."

Ju Xiaofeng had laid special emphasis on the words "completely wiped out" and "not a single trace." If one could compare her speech to a block of wood, one could say that the words "completely wiped out" and "not a single trace" were like knots staring out starkly from the surface of the wood.

I finally got to the root of the problem.

"You must miss your father," I said.

Ju Xiaofeng's eyes reddened, suffused with tears. "No matter how good my step-father, the fact remains—my hair, my teeth, my bones, they are gifts from the man whose memory is completely wiped out of existence. I miss him. But my mother must not know. She would feel guilty, as if she had not done her duty by me. She has done nothing wrong. It is not her fault. The fact is I want to memorialize my father in my own way."

"What do you mean by your own way?"

"I mean to dwell on everything relating to death. For instance, I believe that we each have a soul. I believe that after death a part of us is left behind—otherwise how can we the living mourn the dead?"

I felt in my bones that we were coming to the heart of the matter.

I asked: "How would you identify the part of us that is left behind?"

"Well," she answered, "a piece of land, a flower, a tree... there cannot be a complete blank. That would be too painful for the living."

我说：“所以，你父亲的逝去，让你受不了。所以，你就选择了出售冥位。你希望和你有一样遭遇的人，可以找到寄托自己哀思的地方。其实，你最希望的是知道父亲居住的地方。”

鞠小凤没有任何先兆地放声痛哭。少女的声音清脆而具有穿透力。

鞠小凤的妈妈不顾一切地推开门，想冲进来。我赶忙走出去，好在鞠小凤沉浸在自己的巨大伤感中，并没有发觉这一切。

鞠妈妈焦虑万分地说：“这孩子怎么啦？我拉着她来看心理医生，没想到她号啕痛哭。看样子，旧病未去，新病又来，这孩子是越来越不靠谱了。”

我说：“您放心。她在为自己的父亲感到哀伤。”

鞠妈妈半信半疑说：“她那时候非常小，几乎不记事啊。”

我说：“鞠小凤是个非常聪明敏感的孩子，对父亲的怀念，让她比一般孩子更早熟。这种没有经过处理的哀伤，一直潜伏在她的心灵深处，所以才有了去出卖冥位这样的怪异选择。现在，就让她尽情地哭一场吧。”

我们就这样一直安静地等待着，直到鞠小凤渐渐停止了哭泣。我走进去说：“你可以给你的父亲写一封信，把你所有想和他说的话，都写在里面。”

鞠小凤说：“写好了之后呢？”

我说：“你可以把它放在河流中，也可以系在一棵树上，也可以用火焰烧掉。在古老的习俗中，火焰是通往另一个世界的阶梯。”

鞠小凤擦着眼泪说：“我明白了，冥位其实就在我们思念亲人的任何地方。”

I understood. "So the pain of your father's death is unbearable, he himself having left nothing behind. So you decide to sell burial plots. You want to help people who have lost a loved one to have something to hold on to, something on which to vent their grief. But of course, what is uppermost in your mind is to find your father's resting place..."

Ju Xiaofeng burst out crying, her sobs piercing in an outpouring of grief.

Mrs Ju pushed open the door, trying to enter, but I walked her out. Ju Xiaofeng was so immersed in her own grief that she did not even notice the interruption.

Outside the consulting room, Mrs Ju fretted: "What's going on? I brought her in for counseling, and now she is crying her heart out. The old problem is not yet gone and now we are saddled with a new one. Where is this going to end?"

I told her not to worry, that her daughter is crying for her dad.

Mrs Ju was doubtful. "At the time she was so young; she couldn't have any memory."

I told her, "Ju Xiaofeng is very smart, and very sensitive. Having lost her father, she has matured earlier than most kids. Secretly mourning, she never had a chance to give vent to her grief. This explains her fixation on burial plots. Let her have a good cry."

So we waited quietly for Xiaofeng's sobs to subside. When she quieted down, I walked in and said, "Write a letter to your father. Pour out everything that you have kept pent up so far."

She asked: "And then what?"

"You can let it float in the river or hang it up a tree, or burn it up. According to the ancients, fire is the ladder to the afterlife."

Ju Xiaofeng wiped her eyes, saying, "Now I understand. Burials plots could be any place where we remember the dead."

我忏悔，因为我的身材

女人们对待自己的感情经历，大体上可分为三种。一种是讲，逢人就讲，对熟悉她和不很熟悉她的人，甚至车船旅途中的萍客，都可倾诉。一种是不讲，埋得很深，不少人把它像一种致命的病菌一样，带进坟墓。第三种是通常不讲，但在某一特别的场合和时间下，会对人讲。那种时刻，如果我恰巧成为听众的话，常常生出感动。因为我知道，此时一定有什么特别的情形，痛切地触动了她的内心。我也要感激她对我的信任和这一份特别的缘分。

“那一夜，月亮非常亮。据说是63年以来，月亮最亮的一个晚上。”女孩对我说。

“我是师范院校的学生。读师范的女生，基本上都是家境贫寒的，长相通常也不很好。这样说，我的女同学们，可能会不服气，但我说的是实话，包括我自己，相貌平平。大约读大二的时候，我们就可以做家教了。其实那时，我们和普通大学生所上的课，并没有大的区别，还没学到教学教法什么的，也不一定就能当好如今独生子女的小先生。师范院校的牌子挺能唬人的，再说我们也特需要钱来补贴。

Penance for my Perfect Figure

In dealing with past emotional experiences, women may be divided roughly into three types:

Those that must pour it out—never mind to whom—familiar or unfamiliar faces, chance acquaintances met on trips—they must buttonhole someone to unburden themselves.

The second type of women will never unburden themselves. They will hide their emotional experiences and carry them into the grave.

Then there is a third type, who would not unburden themselves as a rule, but would pour it out on special occasions. Such occasions are usually touched off by something which happened to move the speaker on a soft spot and trigger a spontaneous outpour. I am always very grateful for those special occasions which show the speaker's confidence in me.

It was a moonlit night. According to the girl's information, it was the brightest moon to appear for the last sixty-three years.

She was a student at a teacher's college, she told me. Students at teacher's college are usually plain looking girls from poor families.

"Some of my classmates might not agree," she said, "but I am telling you the truth, and including myself in the picture. Actually our curriculum is not necessarily inferior to the program of a regular college. Though none of us had received any training in teaching skills, yet by the sophomore year, we were supposedly qualified to do some after-class coaching for high school kids. Attendance at teacher's college somehow qualified us. And we needed the money.

“所以，同学们就自己组织起家教‘一条龙’服务，每天派出代表，在大街上支个桌子，上书‘家教’两字，等着上门求助的家长，接了活后再分给大家。谁领到了活儿，会从自己的收入当中，抽一部分给守株待兔的同学——我们称他们为‘教提’。

“有一天，教提对我说，给你分一个大款的女儿，你教不教？我说，钱多不多？他说，官价。我说，你还不跟大款讲讲价？他苦笑着说，讲了，不成。人家门儿清。我说，好吧，官价就官价。他说，那明天下午四点，范先生驾车大门口接你。

“第二天，我提前五分钟到了学校门口。没人。我正好把自己的服装最后检视一遍。牛仔裤、白T恤——挺得体的，既朴素又充满了活力，而且这是我最好的衣服了。

“四点整，一辆我叫不出来名字的红跑车飞驰而来，停在我面前，一位潇洒的中年男人含笑问道：您是黎小姐吗？

“我姓李，他讲话有口音，我也就不计较了，点点头。我说，您是范先生吗？他说，正是。咱们接上头了，快请上车吧，我女儿正在家等你呢。

“我上了车，坐在他身边，车风驰电掣地跑起来。我从来没有坐过如此豪华的车，那感觉真是好极了。他的技术非常娴熟，身上散发着清爽的烟草和皮革混合的气味，好像是猎人加渔夫。总之，很男人。

"So some of us students grouped together and set up business. One of us—our sales representative—'set up office' on a street corner with a sign Home Coaching, waiting for customers to sign up. We who do the teaching pay him a percentage of our earnings.

"One day the sales rep told me that he found a girl whose dad was a big shot.

"I asked how much were they paying.

"He said the usual.

"I said why not try to get more out of him.

"He shrugged and said, 'I tried but the bugger knows the market.'

"I said, 'Alright, regular price.'

"He said, 'Start tomorrow. Mr Fan will pick you up at four.'

"The next day, I got to the gate five minutes ahead of time. No one was there yet and I took the chance to check my appearance: blue jeans and white T-shirt, very appropriate. Low-key, but exuding an air of vitality. It was all I had, anyway.

"At the dot of four, a fancy red sports car drew up beside me. A middle-aged man emerged from the car and asked 'Miss Li?'

"My last name is Li third tone, while the gentleman pronounced Li in the second tone. I assumed that he had a southern accent.

"So I nodded and asked: 'Are you Mr Fan?' He said yes and asked me to get into the car as his daughter was waiting at home.

"I sat in the front seat next to Mr Fan and the car sailed down the street. I had never sat in such a luxurious car. He was a wonderful driver and the ride was exhilarating. Sitting next to me, he exuded a scent of expensive cigarette and fine leather, a combination of hunter and fisherman. Quite a guy, he was."

“他一边开车一边说，他女儿的英语基础不是很好，尤其是胆小，不敢会话。口语的声音弱极了，希望我不要在意。我的目光注视着窗外飞速闪动的街景，不停地点头……心想，同样的建筑，你挤在公共汽车上看，和坐在这样高档的车里看，感受竟有那么大的差别啊。

“很快到了一片‘高尚’住宅区。（我对这个词挺不以为然的，住宅也不是品质，凭什么分高尚和卑下呢？）在一栋欧式小楼面前停下，他为我打开车门时说，我的女儿英语考试成绩每提高一分，我就奖给你一百块钱。

“我充满迷茫地问他，你女儿的英语成绩和我有何相干呢？我是来教历史的。

“那一瞬，我们大眼瞪小眼，然后异口同声地说：对不起，错了。他赶紧带上我，驱车重回校门口，接上那位教英语的黎同学回家，而我找到已经等得很不耐烦的范先生。

“说实话，那天我对范先生的女儿很是心不在焉。这位范先生虽说也是殷实人家，但哪能与那一位范先生相比呢？我心里称那位先入为主的为——范一先生。

“晚上，我失眠了。范一先生的味道，总在我的鼻孔里萦绕。我想，住在那栋小楼里的女人，该是怎样的福气呢？不过，想来素质也不是怎样的好吧？不然，她的女儿为什么那么胆小？要是我有这样的先生和家业，会多么的幸福啊……

"He spoke to me as he drove: his daughter's English not very good, weak at orals, too timid to speak up, hope I did not mind. I kept nodding though my attention was straying to the street scene out the window. Everything looked different being seen through the window of a posh car, rather than through a bus window.

"We soon arrived at an upscale residential area and the car stopped in front of a multi-storied house. As Mr Fan opened the door, he announced: 'I will pay you one hundred yuan for every single digit that my daughter gains in her grades for English.'

"I was mystified: 'What have I to do with your daughter's English? I teach history!'

"We stared at each other, and then said at the same time: 'Sorry, this is a mistake!' He drove me back to the school gate where he deposited me and picked up the girl who was supposed to coach his daughter for English. Meanwhile the other Mr Fan, who had come to fetch me for his daughter's history course, had been waiting in vain.

"I rode back with the 'right' Mr Fan and made some half-hearted efforts to help his daughter in history But my heart was not in it. This was a decent household, it is true, but nothing to compare with the other Mr Fan's establishment. I spent a sleepless night. The fragrance of the first Mr Fan's cigarette lingered in the air around me. I tried to imagine how magnificent it would be to be the lady of that house! Then I thought the lady of Mr Fan's house could hardly be very cultivated, else how could her daughter be so helpless in English? If only I had such a husband and such a house!

“想归想。这年纪的女生，谁没有一肚子的幻想呢？天一亮，我就恢复正常了，谁叫咱是灰姑娘呢！下午四点之前，我又到了校门口，范二先生说好了再来接我。可能是因为头天迟到的缘故，我到的格外早。

“走近校门口，我的心咚咚跳起来——又看到了那辆非凡的红色跑车。我悄悄站在一旁，因为和我没关系。他是来接英语系的黎同学的，这很好理解。

“没想到，那辆红跑车，如水鸟一样无声地滑到了我面前，范一先生温柔地笑着说，李小姐，你好。

“我说，您到得很早啊。

“范一说，昨天我正点到时，你已经到了。所以我想你今天还会到得早，果然不错。我喜欢守时的人，咱们走吧。

“他说着，打开了车门。

“我说，范先生，昨天错了。

“他笑笑说，昨天错了，今天就不能再错。我已将黎同学炒了，重新雇用你。

“我很吃惊，说，你怎么会知道今天我们能见面？

“他说，不要这么惊奇。你惊奇的样子，可爱极了。对于一个商人来说，这点信息有什么难呢？历史系，一个姓氏和‘黎’近似的有着魔鬼身材的女生，现正做着家教……就这样啊。

“我扶着车门说，我不是英语系的。

“他说，你的大学只要是考上的，就可以教我女儿的英语……上车吧，我女儿已经在等了。

"I was engulfed in wishful thinking. Is there a girl who had never ever indulged in daydreams? In the light of day, however, I would go back to normal. Like Cinderella. It is my fate..."

The girl went on with her narrative:

"By four in the afternoon of the next day, I was at the school gate again, waiting to be picked up by the *right* Mr Fan. I made a point of being there early, to make up for my lateness the day before. My heart started pounding when I reached the gate—there it is! The red sports car! I kept myself out of the way, of course—it had nothing to do with me. This Mr Fan has come for the other Miss Li. I knew that perfectly well. However, the red sports car glided to my side.

"Mr Fan smiled as he said: 'How are you, Miss Li?'

"I said: 'You are early' and he said 'Yesterday I arrived on the dot of time but you were already there. So today I expected you to be early too. See, I am right. I like people who keep good time. Let us get moving.' Saying he opened the door.

"I reminded him that yesterday there had been a mistake, and that I coach in history for the daughter of another Mr Fan.

"He smiled and said: 'Yesterday had been a mistake, but not today. I have fired the other Miss Li. Now I am hiring you.'

"'How do you know that I am available today?' I asked him.

"'Don't be shocked—though you are lovely with that shocked look on your face. For someone in business, checking up your data is child's play: History department, surname Li, devil of a figure, part-time coaching...that's about it.'

"'I am not in the English department,' I said as I laid my hand on the handle of the car door.

"'No problem! So long as you have passed the National Test for Higher Education, you can certainly teach English to my daughter. Let's go, she is waiting!'"

“在车上，所有昨天的感觉都复活了。正当我沉浸在速度的快感之中时，范一先生打断了我的美好感受。他说，看来你对自己太不在意了。

“我说，此话怎么讲？

“他说，你穿着和昨天一模一样的衣服。有你这样魔鬼身材的女孩，应该善待自己才是。

“我说，一个穷学生，是无法善待自己的。

“他说，我也当过穷学生，你的处境我能体会。但是，别忘了，你有资源啊。

“我说，我有什么资源啊？芸芸众生而已。

“他说，你的身材非常好，我昨天一眼就被吸引了。一个人，长相好，其实相对来讲比较容易。一张脸，才有多大的面积？对比匀称不算难。就是有些小的瑕疵，比如眼睛不够大，鼻梁不够挺直，做做整容也不难，巴掌大的地方，就那么几组零件，好安排。可一个人的身材，波及全身所有的结构，头颅过大过小都不成，脖子不长不行，脊柱要挺拔，胸腰的比例要适宜，腿更是重中之重，要是短了，纵使闭月羞花也白搭……你呢，刚刚好，所有的搭配都天造地设，你要懂得珍惜啊。而且我提醒你，女性的身材，是很脆弱的结构。上了年纪，就不一样了。锻炼出来的，节食出来的，和天然的，是不一样的……好了，我们到了。

“又是那座小洋楼，但我无心观赏它的精致了。我的心被范一先生的逻辑催动，变得不安分了。这就像一个穷人，守着自己的几亩薄田苦熬。有一天，突然有人对你说，你田里长的那些草，都是人参啊。你还能心平气和吗？

“Once in the car enjoying the speeding, my feelings of yesterday were revived. Mr Fan interrupted my daydreams, saying: ‘Obviously you do not take care of yourself.’

“‘What do you mean?’

“‘You are wearing the same clothes that you did yesterday. With a figure like yours, you should treat yourself better.’

“‘A poor student like me, I can’t afford to treat myself any better.’

“‘I understand, I have been a poor student myself. But don’t forget, you have resources at your disposal.’

“‘What resources? I am just a small potato.’

“‘You have a good figure. I saw it at a glance. Good looks are relatively easy to maintain. A face takes up just that much space; it is easy for the parts to match. Some minor defects—eyes too small, for instance—do not matter much. The bridge of one’s nose, if not straight, can be fixed. The space of a face is so limited, like the palm of my hand, just a few parts, easy to deal with. The trunk of the body, however, which is an entity in itself, is another story. Consider body shape. The head cannot afford to be too big or too small. The neck must be long. The back should be straight. The breasts and the girth must make a perfect match, while the legs can never be too long. If the legs are short. Then it is a hopeless case. In your case, all the shapes and dimensions are heavenly made and just perfect. You must learn to value them. And let me remind you,’ he continued, ‘a woman’s body is a fragile thing. Once age creeps upon you, then it is over. That which is achieved by exercise or by dieting can never pass itself off as the real thing.... Okay, we are there.’

“It was the same Western style building that I saw yesterday, but I was in no mood to observe the details. I was shaken by Mr Fan’s words and felt restless. I was like a poor man who had been content to make a living on his patch of land—until one day someone told him that the grass on his land was actually ginseng! Then his world is turned upside down, and he will lose his peace of mind.

“不过，那天我还是抖擞起精神，辅导范一先生的女儿。我对女主人的羡慕和嫉妒，都不存在了。这是一个没有女主人的家庭，因此那女孩十分孤独内向。她的英语其实不是很差，只是因为不敢说，成绩才糟。

“范一对我很满意，约定以后天天接我来做家教。我说，都是这辆车吗？

“他说，你很在意这辆车吗？

“我说，不是在意，是它美丽。

“他说，我能理解。美丽的东西，人们都想和它在一起。好吧，即使我不能来，我也会派我的司机，开着这辆车来。

“我和范一先生的女儿交了朋友，她的胆子渐渐大起来。嘴一敢张开，成绩就突飞猛进。

“校门口每天准时出现的红色跑车，让我大出风头。有时候下午有课，我就编谎话请假，总之从未误了范一那边。期末，那女孩的英语成绩提高了二十五分，范一递给了我二千五百块钱。

“我就接过来了。心安理得。

“后来，他开始给我买衣服，我不要，他说，我是不忍暴殄天物啊。我就收了……直到有一天，他很神秘地拿出一个纸袋，说是托人特地从国外带回来的时装，送给我。那套衣服漂亮得让人心酸，让人觉得自己以前穿过的都是垃圾。

“你能今天在我家就把这套衣服穿起来，让我看看吗？你知道，我也很爱美丽的东西啊。范一说。

“我本不想答应，但我怕范一不高兴。工钱和奖金，都是我必需的，还有这套华贵的衣服。

"I collected myself and concentrated on helping Mr Fan's daughter over her English. My envy of the mistress of the household evaporated, as there was no mistress. The girl was very introverted. She actually did have a decent mastery of English, but was too timid to speak up, so her grades had suffered.

"Mr Fan was satisfied with my coaching and arranged to pick me up every day. I asked him if he was going to drive the same car. He wondered if I had a special feeling for the car. I replied that the ride was not the issue—it was just that the car was so beautiful. Mr Fan said that he understood. Everybody loves beautiful things, he added. He promised that on occasions when he could not drive, he would send his chauffeur with the car.

"I became friends with Mr Fan's daughter. She gradually overcame her shyness and started to speak English, and her grades shot up.

"The red sports car that showed up at the school gate every afternoon got me a lot of attention. On occasion, I would make up an excuse and miss a class myself, but I never missed a single coaching session for Mr Fan's daughter. At the end of the term, the girl's grade shot up by 25 marks, and Mr Fan gave me 2500 yuan, just as he had promised. I accepted the money matter-of-factly. I deserved it.

"Then Mr Fan started to buy me clothes. I refused at first. But he insisted, saying that otherwise it would be wasted, so I accepted—until one day he took out a package with a great air of mystery, saying that it was a gift that he had asked a friend to bring back from abroad, and the outfit was so exquisite that all other clothes were rags in comparison.

"'Could you try it on right away,' Mr Fan suggested, so that he could take a look. 'I love beautiful things too,' he added. I really felt reluctant, but then I was concerned that Mr Fan might be offended. I called to mind my fees for coaching and the extra bonus, which I really depended on to make ends meet—and now this special gift.

“我把卫生间里面门上的小疙瘩按死，开始换衣服。正当我把旧衣服脱下，新衣服还没上身的时候，门无声无息地开了。

“我想看看自己的眼光，对你的三围的估计准不准？范一说。

“我呼救反抗……偌大的房间里，只有我们两人，女孩到同学家去了。暴行之后，范一扔下一笔钱，说，我是很公平的。你们做家教，是按小时收钱，明码标价。我也是。你的每一公分胸围，我付一笔钱。你的腰围比臀围每少一公分，我付一笔钱。我可以告诉你，我从来没有给过任何一个小姐这么多的钱。你真是魔鬼身材啊。

“我很想到公安局告他，可我怕舆论。每天招摇的红跑车，让我气馁。我也很想把钱扔到他脸上，然后扬长而去。那是电影里常常出现的镜头，但是，我做不到。我缺钱，我已经付出了高昂的代价，我要为自己保存一点物质补偿。

“我想，一个人是不是记得住那些惨痛的教训，不在于片刻的决绝，更在于深刻的反省吧。

“我再也没有见过范一。有时候，在镜子面前欣赏自己优美的身材的时候，我会想起范一的话。我承认这是一种资源。但是，所有的资源，都需要保护。越是美好的资源，越要珍惜。女人，最该捍卫的，不就是我们的尊严吗？！”

在明月的照耀下，我看到她脸上的清泪。

"I went to the lavatory, closed the door, turned on the lock and started undressing. Just as I was out of my own clothes and had not yet put on the new dress, the door knob turned silently and the door opened. Mr Fan said he wanted to check if he had got my three critical measurements right when ordering the dress.

"I cried out in alarm. His daughter had left to visit with a friend, so there were just the two of us in that large house. He violated me.

"Then he threw down a stack of banknotes, saying 'I play fair. You charge by the hour for coaching. And I have my standards. Now for every inch of your bust, there is a price. And for every inch of the difference between your waist line and your butt measurement, there is an added prize! Let me tell you I have never paid so much for any other woman. Your body measurements are absolutely fiendish!' Saying, he threw down the money and left.

"My first instinct had been to report him to the police. But I was scared of the publicity. I had also wanted to fling the money in his face and stride out of the room, as they do in the movies. But I could not bring myself to do it. I needed the money. I have already paid a heavy price. I wanted some compensation. I will never forget this deadly lesson. No need for hysterics. I need to think back and draw the lesson of a lifetime.

"I never saw Mr Fan again. Sometimes as I look at the mirror and admire my own outline, his words would come back to me. A woman's body, I acknowledge, is a resource. All resources should be protected. The more precious the resource, the more they need protection. But for women, isn't personal dignity our most precious resource?"

Under the flickering moonlight, I saw the tears on her cheeks.

我忏悔，

因为我的身材

Penance for my Perfect Figure

你的悲伤不是全世界都要知道，要学会一个人担当。

最重的咨询者

我猜你第一眼看到这个题目，一定以为是“最重要的咨询者”。很抱歉，不是最重要，是最重。你可能要大吃一惊，说你们的心理咨询室里还设磅秤吗？每个来咨询的客人，都要量体重吗？

并没有人体秤，我也从来没有问过来访者的体重。只是这位来访者实在太胖了，不用任何器械，我也能断定他在我所接待过的来访者中体重第一。

他穿了一条肥大的牛仔裤，一看就是那种出口转内销的外贸尾单货，专供欧美等国特大号胖子装备的。上身是一件黄蓝相间的花衬衣，有点儿苏格兰格子的味道，想来是从国外淘买回来的，亚洲人难得有这样庞大的规格。他名叫武威，正在上大学三年级。

“我好着呢！什么毛病也没有！”武威开门见山地说。他小山似的身体将咨询室的沙发挤得满满当当，腰腹部的赘肉从沙发的扶手镂空处挤出来，好像是脂肪的河流发山洪外溢了河道。我暗自庆幸当年置办办公家具的时候，选择了不锈钢腿的沙发。若是全木质精雕细刻的，在这样的负荷之下，难免断裂。

我说：“既然您觉得自己一切正常，为什么到我们这里来呢？”

我问这话，不单单是一个询问策略，实实在在也是自己心中的困惑。当然了，武威的体型令人瞠目结舌，但如果当事人不觉得这是一个问题，心理师也犯不上自告奋勇迫不及待地为人排忧解难。

A Very Weighty Problem

At first glance, you would assume that I am talking about a very important visitor. Wrong. I am talking about someone who is literally weighty. You might wonder—is there a weight in your office? Do people check their body weight before walking in for counseling?

Actually there is no weight in my office and I have never concerned myself about the body weight of my visitors. But this visitor was so palpably over-weight. I did not need to check his weight to see that he was the most over-weight person that had ever walked into my office.

He was in a pair of jeans, the baggy kind designed for over-weight people in the West. He wore a blue-and-yellow checkered shirt, the casual style popular at sales abroad. His size is rare among Asians. He is called Wu Wei, and is currently in his third year in college.

"I'm fine; no complaint of any kind," announced Wu Wei as he squeezed himself into the sofa, his fleshy hips peeping through the checkered handsets on each side. I secretly congratulated myself that I had ordered metal support for the sofa—the carved wood variety would have caved in under Wu Wei's weight.

I started the consultation by saying, "If there is nothing wrong with you, why are you here?"

I asked this question, not as a strategy to start a conversation. I was really perplexed at his coming. His body weight was indeed a problem, but if he did not mind, it is not the counselor's job to initiate any weight loss program.

武威一笑，笑容有一种孩子般的天真。他说：“我说我觉得自己正常，但这并不代表着我的家人也觉得我正常。”

我说：“这么说，是家里人让你来看心理医生的？”

武威说：“可不是嘛！他们说我太胖了，马上就要面临大学毕业找工作，像我这样的体型，会受到歧视，更甭说以后找对象结婚的事了。总之，他们让我减肥，我吃过各式各样的减肥药，喝过名目繁多的减肥茶，还尝试过针灸推拿揉肚子……”

我问：“什么叫揉肚子？”

武威说：“一种新流行起来的减肥方法，就是好几个人在你的肚子上像和面一样揉啊揉的，据说能把腹部的脂肪颗粒粉碎，这样就可以排出体外了。还有一种吸油纸，就像胶布一样贴在你想减肥的部位，大概过上一个小时，就会看到那片纸变透明了，全都是油滴。”

我大吃一惊，以我当过20年医生的经验，绝对不相信人体内的脂肪会被一张纸榨出来。

“这是真的吗？”我问。

武威说：“有一次，我把吸油纸贴在冰箱外壳上。一个小时之后，吸油纸也是油光闪闪的。”

我愤然，怎么能这样骗人！

武威说：“现在社会上流行以瘦为美，商家就利用人们的这种心理，大发减肥财呗。”

我发现武威虽然看起来动作迟缓，但思维清晰敏捷。

我说：“想必你尝试过种种减肥方法，都没效果。”

Wu Wei smiled, an innocent happy smile, like that of a child. "When I say that I feel normal, it doesn't mean that my family agrees."

"So your family advised you to get counseling?"

"Exactly! They said that I am too fat, that I will be discriminated against when I start looking for a job, not to mention dating and marriage. They want me to slim down. I have taken all kinds of pills and tried all kinds of drinks for weight loss. I have even tried acupuncture and belly massage, but nothing worked."

"What do you mean by belly massage?"

"It is a new treatment for weight loss, with several people massaging your belly. The idea is to break up the molecules of fat and discharge them from the body. There is another kind of weight loss, using fat-absorbing paper spread over your tummy. After an hour or so when the paper is lifted, it is soaked with your body fat."

I was shocked. In all my twenty years of practice, I had never heard of human fat being soaked up by a piece of paper. "Is it true?" I asked. Wu Wei said that once he left a piece of that kind of paper over the fridge, and after a while the paper was soaked with shiny fat.

"So it was a hoax," I exclaimed.

"Well," said Wu Wei, "the standard for beauty is thinness, so business people play on your weakness, and make money through bogus weight loss programs."

I discovered that Wu Wei's body movements were slow, but that his mind was nimble.

I said: "So you had tried all kinds of weight loss programs with no effect?"

武威说："您说对了一半。就我尝试过的方法，公平地说，除了吸油纸是彻头彻尾的骗术以外，其他的多少都有一些效果。它们之中要么是用了泻药，要么使用了西药抑制人的食欲，每次我都能成功地减肥几十公斤。"

我又一次坠入雾海。若是每一次都减肥成功，那么武威目前就不会是如此庞然大物了。或者说，他以前简直重如泰山？

看到我百思不得其解的模样，武威说："是的，每一次都成功，可是，您知道反弹吗？"

我说："知道，就是体重又恢复到原来的分量了。"

武威说："岂止是原来的分量，更上一层楼了。我就这样，一次又一次地减肥，然后一次又一次地比原来更肥。"

我觉得武威说完这句话，应该愁眉苦脸，起码也会叹一口气吧？可是，武威依然是安之若素的模样，甚至嘴角还浮现出隐隐的笑意。

我有点儿怀疑自己的眼神，但是，没错，武威脸上并没有任何沮丧的神气，看来，他说自己没有问题，也不是毫无根据的。但是，面对着这种明显不正常的体重，还要说一切正常，这是不是正是要害所在呢？

我对武威说："我看你对自己的体重，并不觉得有什么不合适的地方。"

武威好像遇到了知音，说："哎呀，您可真说到我的心里了，我并不觉得这不正常。"

把一个明显不对头的事，说成正常，这也是问题啊。我说："武威，你可以有一个选择。你要是觉得自己没有一点儿问题，就可以走了。你要是希望自己变得更好，咱们就来探讨一下有关问题。毕竟，你的体重超标了，这是一个事实。"

"To be fair," Wu Wei said, "you are half right. Apart from the hoax of fat-absorbing paper, other methods that I had used did have some effect: some loosened your bowls; others suppressed your appetite. They usually help me lose several dozen kilos at a time."

I was again confused. "If you had managed to lose several dozen kilos through each session, you should have lost considerable weight by now..."

Seeing my perplexity, Wu Wei said, "True, there was some measure of success each time, but have you heard of the term *rebound*?"

"Of course," I said, "your weight is back to what it had been."

"Back to what it had been plus something over! That is what I have been going through. Losing weight and gaining it back, losing and gaining..."

I had expected Wu Wei to be downcast as he recounted his plight, or at least give a sigh. Actually, Wu Wei recounted the narrative quite matter-of-factly, even with a wan smile hanging on his lips.

I could hardly believe my eyes, but it was true. Wu Wei did not seem a bit downcast. When he said that he did not have a problem, he knew what he was talking about. On the other hand, to claim that he was alright when he was so obviously overweight—isn't that a problem in itself?

I said: "It seems that your weight does not bother you..."

"You have really laid your finger on the heart of the matter," he exclaimed. "I don't consider my weight a problem."

Now, to be so glaringly overweight and think nothing of it, that is definitely a problem in itself. I said, "Wu Wei, you may think you don't have a problem, in which case, you are free to leave. Or you may wish to improve your condition, in which case, we can discuss the issues involved. After what is said and done, you are overweight, and that is a fact."

武威迟疑了一下。看来，他是一个好脾气的胖子，所以，他并不想忤逆父母的意愿，就乖乖地来见心理医生了。不过，他打算走个过场，然后就照样我行我素。现在，面临选择，他费了思量。过了一会儿，他说：“你说这话我愿意听——谁不愿意把自己变得更好呢？我愿意和你讨论一下我的体重问题。”

很好，显著的进步。武威终于承认自己的体重是一个问题了。

我说：“你从小就比较胖吗？”

武威连连摇头说：“我小的时候一点儿都不胖。从12岁零3 个月的时候，开始发胖。以后就越发不可控制，差不多每年长20斤。要说一个月长一斤多肉，也不是什么了不起的事，但日积月累，就成了现在的样子。”

这段话初听起来，好像很普通。但我注意到了一个奇怪的数字，12岁零3个月。按说体重增加并不是突然发生的，但武威为什么把日期记得那样清楚呢？

我说：“武威，当你12岁零3个月的时候，发生了什么？”

武威低下头说：“我不能告诉你。”

我说：“为什么？”

武威说：“因为一想起那段日子，我太悲伤了。”

我说：“武威，将近10年过去了，你还这样痛苦。我猜想，这也许和你的一位挚爱的人离去有关。”

Wu Wei stood, undecided. Obviously he is a good-natured fellow, reluctant to go against his parents' wishes. So he came for a consultation, but with no intention of doing anything about his weight. Now, faced with a real choice, he demurred for a while, then coming to a decision. He announced, "I like what you said. I would like to improve my condition and do something about my weight. Let's talk."

Good. Now we are getting somewhere—Wu Wei acknowledges that he has a weight problem.

"Were you overweight since childhood?"

Wu Wei shook his head vigorously. "I started putting on weight when I was twelve years and three months old. From then on, it went out of control. I would put on twenty jin per year. Actually gaining one jin or so per month is no big deal, but to keep growing at that rate, you see what I am now." What he was saying sounded quite off-hand, but one remark caught my attention—"twelve years and three months old."

I asked: "Wu Wei, what happened when you were twelve years and three months old?"

Wu Wei bent down his head: "I can't tell you."

"Why?" I asked.

"Thinking back to those days is very painful."

"Wu Wei," I said, "Ten years have passed since then and you are still hurting. Perhaps it has to do with the death of someone whom you love deeply?"

武威抬起头来，我看到他的眼珠被泪水包裹。他说：“您说对了。我从小就是和外婆在一起，她是个非常慈祥的老太太。我从她那里，得到了温暖和做人的道理。我觉得她这样的好人，是永远不会死的。可是，她得了癌症。很多人得了癌症，也都可以治疗，比如化疗什么的，就算不能挽回生命，坚持个三年五载的也大有人在。可我外婆什么治疗都不能做，从发现患病到去世，只有短短的20天。我痛不欲生，于是就拼命吃饭，从那以后，就踏上了变胖的不归路……”

我的脑海开始快速运转。按说痛不欲生的结果，是令人食欲大减，饭不思茶不饮的，似这般暴饮暴食胡吃海塞搞得体重骤升的，实在罕见。

我说：“原谅我问得可能比较细，你吃下那么多东西的时候，想的是什么？”

武威说：“我想这就是纪念我外婆的一种方式。”

我又一次糊涂了。祭奠亲人的方式，可能有千千万万种，但用超常的食欲来思念外婆，这里面有着怎样的逻辑？

我说：“你外婆一直鼓励你多吃饭吗？”

武威说：“没有。外婆是非常清秀的江南女子，直到那么老的年纪，都非常美丽，每餐只吃一点点饭。”

我说：“那么，你为什么要用吃饭悼念外婆呢？”

武威陷入了痛苦的回忆。许久，他喃喃地说：“也许……是因为……我听到了一句话。”

我说：“那是一句怎样的话？”

Wu Wei lifted his head and I saw his eyes suffused with tears. "You are right," he said, "I had been brought up by my maternal grandma since childhood. She was such a kind old lady. She wrapped me in her loving care and taught me the principles of life. I had assumed that such good people like my grandma would live for ever. And then she had cancer. Many people have cancer and they are cured—chemotherapy and stuff. Even if they are not cured, they could hold out for a couple of years. But my grandma's cancer could not be cured. She died within twenty days of being diagnosed. I crumbled up, losing interest in everything. Just ate non-stop. That is how I have become what I am now. And there's no turning back."

My head was spinning. When one is devastated by grief, one has no appetite for either food or drink. To gain so much weight through reckless eating as a reaction to disaster is totally unheard of.

"Pardon me for asking," I said, "but what were you thinking of while you were gorging?"

"To me, it was a way of memorializing my grandma."

I was totally mystified. There are so many different ways of remembering the dead, but not by gorging. It did not make sense to me at all.

"Was your grandma in the habit of encouraging you to gorge yourself?"

"Oh no," he said, "Grandma had been a delicate southern beauty. She just nibbles a little over meals."

"Then why are you remembering her by gorging yourself?"

For a while, Wu Wei seemed lost in painful memories. After a while, he murmured, "Perhaps it was because of something I overheard..."

"Overheard?" I pursued.

武威用手支撑着巨大的头颅，说："那一天，我到医院去看望外婆。正是中午，大家都休息了。当我路过医生值班室的时候，听到两位值班医生在说话。男医生说，13床的治疗方案最后确定了没有？女医生说，没有什么治疗方案了，就是保守对症，减轻病人一点儿痛苦。男医生问，干吗不手术呢？女医生答，年纪太大了，如果手术，很可能就下不了台子，比不做还糟糕。男医生又开言，那么化疗呢？资料上说，现在新的药物对这种癌症效果不错的。女医生接着回答，13床太瘦弱了，化疗方案一上去，人肯定就不行了，还不如这样熬着，活一天算一天……

"13床，就是我的外婆啊。

"医生们的这段对话，让我留下了非常深刻的印象。我觉得外婆的死，就是因为她太瘦了，瘦到无法接受治疗，如果她胖一点儿，就能够战胜死神，就能一直陪伴在我身边……"

武威断断续续地讲着，他的眼泪一滴滴洒落在黄绿相间的格子衬衣上，让黄的地方更黄，绿的地方更绿。胖人的泪滴也比一般人的要硕大很多，每一滴都像一颗透明的弹球。

Holding up his head with one fleshy hand, Wu Wei started to tell his story. "I went to the hospital to see Grandma. It was the noon break for napping. I passed by the staff room and heard two doctors chatting. A male voice said:

"'Have you finalized the treatment plan for Bed 13?'

"'Not yet,' a woman's voice answered, 'there's not much to be done, seeing how far gone she is...'

"'What about operation?' the male voice asked.

"'She's too old. She might never get up from an operation. Just more suffering.'

"The male voice persisted: 'What about chemotherapy? It could be quite effective.'

"'Not for Bed 13. She's too thin, too weak. She will not survive it. Better this way. Let her live out the days...' "

"Bed 13! That was my Grandma! The words of the two doctors shocked me badly. I was convinced that my Grandma had died because she had been too thin, too weak to undergo any treatment. If she had been more filled out, she could have overcome Death. She would still be with me."

Wu Wei wept as he told me the story. It seemed that even the tears of overweight people were bigger than the usual, each single tear of Wu Wei's round and transparent, dripping on his blue-and-yellow checkered shirt.

我默默地坐着，能够想见至亲的人离去，给当年的小男孩以怎样摧毁般的打击。他以自己的方式表示着痛入心肺的哀伤，表示着对于死神的强大愤怒，表示着对于外婆的无比眷念……难怪他不认为这是不正常的，难怪他在每一次减肥之后都让自己的体重增加。

在接下来的多次咨询中，我和武威慢慢地讨论着这些。当然，我不能把自己的判断一股脑儿地告知他，而是在我们的共同探讨中，渐渐向前。

武威后来成功地减下了50公斤体重，成了英俊潇洒的靓仔，对外婆的悼念也化成了力量，他各方面都很优秀。

I remained sitting, musing on the shock and pain that a sudden death would inflict on a little boy. He expressed his sorrow in his own way, expressed his anger at death in his own way, memorializing his Grandma in his own way. Now it makes sense that he did not see his weight as a problem. No wonder that after every weight loss he would let his weight bounce back.

Over the following sessions we sat down, Wu Wei and I, and we went over the issues. I never imposed my own views on him. We just moved slowly forward through the sessions.

Wu Wei ultimately succeeded in losing fifty kilos and turned out a handsome young man. The memory of his Grandma became for him a drive to make life more meaningful.

失眠失恋，你究竟失去了什么

一个身材高大的男青年倚在一个瘦弱的女子身上，踉踉跄跄地走进心理咨询中心。工作人员以为他患了重病，忙说："我们这里主要是解决心理问题的，如果是身体上的病，您还得到专科医院去看。"

女子搀扶着男青年坐在沙发上，气喘吁吁地说："他叫瞿杰，是我弟弟。我们刚从专科医院出来，从头发梢到脚后跟，检查了个底儿掉，什么毛病都没查出来。可他就是睡不着觉，连着10天了，每天24小时，什么时候看他，他都睁着眼，死盯着天花板，任啥话也不说。各种安眠药都试过了，丝毫用处都没有。再这样下去，就算什么病也不沾，人也会活活熬死。专科医院的大夫也没辙了，让我们来看心理咨询。求求你们伸出援手，救救我弟弟吧！"

姐姐涕泪交流，瞿杰仿佛木乃伊，空洞的目光凝视着墙上的一个油墨点，无声无息。

瞿杰进了咨询室，双手拄着头，眉锁一线，表情十分痛苦。

我说："睡不着觉的滋味非常难受，医学家研究过，一个人如果连续一周不睡觉，精神就会崩溃，离死亡就不远了。"

"你以为是我不愿意睡觉吗？你以为一个人想睡就睡得着吗？你以为我失眠是我的责任吗？你以为我就不知道人总是睡不着觉就会死的吗？"瞿杰突然咆哮起来，用拳头使劲击打着墙壁，因为过分用力，他的指节先是变得惨白，继而充血发暗，好像箍着紫铜的指环。

Comparisons are Odious

A hulk of a fellow, leaning on the shoulder of a slender young woman, struggled into the clinic. A staff member went forward and advised them to go to a regular hospital, as ours was a counseling center for psychological problems.

The young fellow lurched into a chair, huffing and puffing, The young woman explained: "He is my brother Zhai Jie. We have just come from a hospital. The doctors had examined him from top to toe and could find nothing wrong with him. But he can't fall asleep; it is ten days by now. He keeps staring at the ceiling without saying a word. He has tried all kinds of sleeping pills, which didn't help. He will waste away if this goes on. The doctor could do nothing for him and advised us to come for counseling. Please be kind and help my brother." She was deluged in tears, but her brother remained unmoved and unmoving, staring at nothing.

The young man was helped into the consulting room, eyebrows knotted, his two hands holding his head, a picture of pain.

I said: "Sleep deprivation is very painful. According to medical research, if a person is deprived of sleep for a whole week, he or she might collapse, and death is just around the corner."

"Do you suppose that I don't want to sleep? Do you think one can fall asleep at a moment's notice? Do you think that insomnia is my own choice? As if I didn't know that lengthy deprivation of sleep could lead to death!" Zhai Jie had a sudden outburst, shouting and banging at the wall until his fingernails, squeezed against the wall, turned white and then purple.

我平静地看着他，并不拦阻。他需要一个发泄，虽然我暂时还不知道导致他强烈失眠和情绪的原因是什么，但他能够如此激烈地表达情绪，较之默默不语就是一个进步。燃烧的怒火比闷在心里的阴霾发酵成邪恶的能量，好过千倍。

至于他把怒火转嫁到我身上，我一点也不生气。虽然他的手指指点的是我，唾沫星子几乎溅到我脸上，指名道姓用的是“你”，似乎我就是令他肝胆俱碎的仇家，但我知道，这是情绪的渲染和转移，并非和咨询师个人不共戴天。

一番歇斯底里的发作之后，瞿杰稍微安静了一点。

我说：“你如此憎恨失眠，一定希望能早早逃脱失眠的魔爪。”

他翻翻黯淡无光的眼珠子说：“这还用你说吗！”

我说：“那咱们俩就是一条战壕的战友了，我也不希望失眠害死你。”

瞿杰说：“失眠是一个人的事情，你就是愿意帮助我，又有什么用！”

我说：“我可以帮你找找原因啊。”

瞿杰抬起头，挑衅地说：“好啊，你既然说要帮我，那你就说说我失眠到底是什么原因吧！”

我又好气又好笑，说：“你失眠的原因只有你自己知道，你要是不愿意说，谁都束手无策。要知道，失眠的是你而不是我。你若是找不到原因，或是找到了原因也不说，把那个原因像个宝贝似的藏在心里，那它就真的成了一个魔鬼，为非作歹地害你，直到害死你，别人也爱莫能助，无法帮到你。”

瞿杰苦恼万分地说：“不是我不说，是我真的不知道为什么失眠。”

我说：“你失眠多长时间了？”

瞿杰说：“10天。”

I watched him on the side without doing anything to stop his outburst. He needed to let it out.

I had no clue as to the cause of his depression and sleep deprivation, but I was sure that this outburst was better than mute suppression. It is much better to release the flame of resentment, rather than let it simmer in the heart of darkness and end in a deadly outbreak. As to his taking out his anger on me, it did not affect me at all. He pointed his finger at me, spluttering in fury, behaving as if I was his worst enemy. I knew that it was nothing personal, just a way of nailing down a target and letting off steam.

As Zhai Jie took a little break after his tirade, I got a word in: "Insomnia is such a nuisance; I am sure you want to get out of its clutches."

"That goes without saying," he barked, rolling his bleary eyeballs.

"Then we are on the same side," I added, "I don't want insomnia to be the death of you."

"Insomnia is a personal thing—even if you want to help, there is nothing you can do about it!"

"Well, I can help you find out the reason behind it!"

"Good!" Zhai Jie looked up challengingly, "Since you want to help, why don't you start by telling me the reason for my insomnia!"

I didn't know whether to laugh or to scold. I said: "Only you yourself know the reason why you can't go to sleep. If you really don't know, or knowing, want to keep it to yourself, no one can do anything about it. But let me tell you this—if you knowingly keep it to yourself like a secret treasure, it will turn into a demon to torment you—all the way to your grave! And no one can help!"

Zhai Jie, a picture of dejection, said earnestly: "I don't mind talking about it, but I really and truly don't know what caused my insomnia."

"How long have you been sleepless?"

"Ten days."

我说："在失眠的时候，你想些什么？"

瞿杰说："什么都不想。"

我说："人的脑海是十分活跃的，只要我们不在睡眠当中，我们就会有很多想法。你说你失眠却好像什么都不想，这很可能是因为有一件事让你非常痛苦，你不敢去想。"

瞿杰有片刻挺直了身子，马上又委顿下去，说："你是有两下子，比那些透视的X光共振的核磁什么的要高明一点。他们不知道我脑子里想的是什么，你猜到了。我承认你说得对，是有一件事发生过……我不愿意再去想它，我要逃开，我要躲避。我只有命令自己不想，但是，大脑不是一个好的士兵，它不服从命令，你越说不想它越要想，这件事就像河里的死尸，不停地浮现出来。"

"我只有一个笨办法，就是用其他的事来打岔，飞快地从一件事逃到另外一件事，好像疯狂蔓延的水草，就能把死尸遮挡住了。这法子刚开始还有用，后来水草泛滥成灾，死尸是看不到了，但脑子无法停顿，各种各样的念头在翻滚缠绕，我没有一时一刻能够得到安宁，好像是什么都在想，又像是什么都不想，一片空白……"说到这里，他开始用力捶击脑袋，发出空面袋子的噗噗声。

我表面上镇静，心里还是有点担心，怕这种针对自我的暴力弄伤了他的身体，做好了随时干预的准备。

过了一会儿，他打累了，停下来，呼呼喘着粗气。我说："你对抗失眠的办法就是驱动自己不停地想其他的事情，以逃避那件事情。结果是脑子进入了高速运转的状态，再也停不下来。你现在能告诉我那件让你如此痛苦不堪的事情，究竟是什么吗？"

他迟疑着，说："我不能说。那是一个妖精，我好不容易才用五花八门的事情把它挡在门外，你让我说，岂不是又把它召回来了吗？"

"What were you thinking of when you were lying in bed, sleepless?"

"Nothing."

"The human mind is very active. As long as you are not asleep, there are always thoughts floating in your mind. In your case, if you insist that you were not thinking of anything while sleepless, it means that there is something that you want to drive out of your mind!"

Zhai Jie stiffened at my words, then slumped back into his chair. "Well, you did have a trick or two up your sleeve, better than those X-ray contraptions. They have no clue as to what is in my head—you have nailed it. Yes, something happened—I don't want to think about it, I want to leave it behind me, I order myself to stop thinking about it. But somehow my mind does not take my orders; the more I want to drive away the thought, the more I brood on it. Like a dead body in the river, it keeps coming up to the surface. My only resort is to drive the thought away with something else, like using seaweeds to cover up a floating corpse. It worked for a while. The corpse had disappeared, but my mind could not stop spinning. My mind became a complete mess, one moment filled to bursting, the next moment a complete blank." Saying which, Zhai Jie started hitting his head with his knuckles.

I tried to keep calm though I was worried that he might hurt himself by this kind of self-punishment. I prepared myself to help him as soon as he was ready. Gradually he got tired and stopped the violence, huffing and puffing. I said to him: "Now it is clear—the reason for your insomnia is that you keep thinking nonstop of other things to keep away the thought of that particular problem, and the effort has set your mind spinning out of control. And now, will you tell me that unmentionable thing that made you suffer so much?"

After a slight hesitation, Zhai Jie said: "No, I cannot talk about it. It is a horror. I have finally managed to get rid of it by soaking myself in other concerns. Talking about it will only bring it back."

我说："我很能理解你的恐惧，也相信你让自己的大脑，不停地从一个问题跳到另外一个问题，用飞速运转抗拒恐怖。在最初的阶段，这个没有法子的法子，在短时间内帮助过你，让你暂时与痛苦隔绝。但是，随着时间的延续，这个以折磨取胜的法子渐渐失灵了。你变得疲惫不堪，脑子也没办法进行正常的思维和休息，你就进入了混乱和崩溃，这个法子最终伤害了你……"

瞿杰好像把这番话听了进去，用手撕扯着头发。我不想把气氛搞得太压抑，就开了个玩笑说："依我看啊，你是饮鸩止渴。"

瞿杰好奇地问："鸩是什么？渴是什么？"

我说："渴就是你所遭遇到的那件可怕的事情。鸩就是你的应对方法。如今看来，渴还没能把你搞垮，鸩就要让你崩溃了。渴是要止住的，只是不能靠饮鸩。我们能不能再寻找更有效的法子呢？况且直到现在，你还那么害怕这件事卷土重来，说明渴并没有真正远离你，鸩并没有真正地救了你。如果把这个可怕的事件比作一只野兽，它正潜伏在你的门外，伺机夺门而入，最终吞噬你。"

瞿杰的身体直往后退缩，好像要逃避那只野兽。我握住他的手，给他一点力量。他渐渐把身体挺直，若有所思地说："您的意思是我们只有把野兽杀死，才能脱离苦海，而不是只靠点起火把敲响瓶瓶罐罐地把它赶走？"

我说："瞿杰你说得非常对。现在，你能告诉我那只让你非常恐惧的野兽是什么吗？"

瞿杰又开始迟疑，沉默了漫长的时间。我耐心地等待着他。我知道这种看起来的沉默，像表面波澜不惊的深潭，水面下风云变幻，正进行着激烈的思想斗争——说还是不说？

"I understand the hole you are in," I said, "I understand how you make your mind jump from one topic to another, letting your mind spin non-stop, to shield yourself against the thought of that horror. At the time it was a last resort, and it did shield you from pain for a while. But with the passing of time, this kind of self-inflicted punishment loses its efficacy. You get yourself exhausted; your mind goes blank and you mess up your own life. You end up being the injured party."

My point seemed to have sunk in, as Zhai Jie sat unmoving, pulling at his hair. I tried to lighten the atmosphere, saying: "To me it seems as if you are 'drinking poison to quench your thirst.'" Zhai Jie picked up his ears: "What poison? What thirst?" He was curious.

I explained: "The thirst is the problem—whatever it is—that had landed in your lap. The poisonous drink is your way of dealing with it. So far as I can tell, the thirst had not affected you much, but the poisonous drink is dragging you down. The thirst should be quenched, but not by drinking poison. Suppose we look for better ways to deal with your problem? Especially as it is obvious that the thirst has not left you and the poisonous drink is still around. If we compare your problem to a beast of prey, we could say that it is just outside your door, waiting to spring on you." Zhai Jie backed away, as if to escape from that imaginary beast of prey. I held his hand in support. He gradually straightened himself up saying,"You mean that we should kill the beast once and for all, not bang on pots and pans to frighten it away."

"You got it, Zhai Jie. And now can you tell me the name of the monster that had so terrified you?" Zhai Jie was lost again, keeping his thoughts to himself. I waited patiently. I knew that this kind of silence was like a pool of water, placid on the surface, but turmoil down beneath—to speak or not to speak?

终于，瞿杰张开了嘴巴，舔着干燥的嘴唇说：“我……失……恋了。”

原本我以为让一个英俊青年如此痛不欲生的理由，一定惊世骇俗，不想却是十分常见的失恋，一时觉得小题大做。但我很快调整了自己的思绪，认真回应他的痛楚。心理问题就是这样奇妙，事无大小，全在一心感受。

任何事件都可能导致当事人极端的困惑和苦恼，咨询师不能一厢情愿地把某些事看得重于泰山，而轻视另外一些事情，以为轻若鸿毛。唯有当事人的情绪和感受，才是最重要的风向标。

我点点头，说：“谢谢你对我的信任。失恋的确是非常令人痛苦的事情，有时候足以让我们颠覆，怀疑整个世界。”

瞿杰说：“我没有把这件事告诉任何人。”

我说：“你不说，一定有你不说的理由。”

瞿杰说：“没想到你这样理解我。你知道我为什么不说吗？”

我老老实实地回答：“不知道。如果你告诉了我，我就知道了。”

瞿杰说：“你看我条件如何？”

我说：“你指的条件包括哪些方面的呢？”

瞿杰说：“就是谈恋爱的条件啊。”

我说：“每一代人都有每一代人的条件，我的眼光可能比较古旧了，说的不对供你参考。依我看来，你的条件不错啊。”

瞿杰第一次露出了笑容说：“岂止是不错，简直就是优等啊。你看我，一米八三的高度，校篮球队的中锋，卡拉OK拿过名次，功课也不错，而且家境也很好，连结婚用的房子家里都提前准备了……”

Finally, Zhai Jie stuck out his tongue to lick his cracked lips and muttered: "I—have been—disappointed in love." Disappointed in love! I had assumed that it would have taken something out of the ordinary to have broken the back of such a handsome young man, and it turned out to be so banal. A storm in a tea cup! I quickly collected my thoughts and settled down to deal with his pain. It is so typical of psychological problems. There are no fixed standards—big or small, it is all up to the individual's personal reaction. Anything could lead to a crisis. The psychological counselor cannot risk making a mountain of one case and dismiss another as a molehill. It is only the personal mood and feelings of the patient that determines the gravity of the problem. I nodded my head as I murmured: "Thank you for trusting me. To be disappointed in love is indeed a painful experience. Sometimes it would seem as if the world is reeling..."

Zhai Jie said: "I have never told this to anybody..."

"You must have your reasons."

"Thank you for understanding. Do you know my reason?"

"Of course I don't! Unless you tell me."

He changed the subject by asking: "What do you think of my qualifications?"

"Your qualifications? For what?"

"I mean qualifications for getting a girlfriend."

"Every generation has its own standards for falling in love. My standards are from an older generation. You are welcome to point out where it is obsolete. But according to my standards, you are not bad."

For the first time since his arrival, I saw him smile. "Well, I certainly hope I am something above *not bad*. I would say I am highly qualified. Look at my height—1.83 meters. I play center in the college basketball team. I won prizes at karoke singing contests. My grades are fine, and my family circumstances are above average—even a new house for my wedding is ready and waiting."

我说："万事俱备只欠东风了。"

瞿杰说："是啊，这个东风就是一位女朋友。"

我说："你的女朋友究竟是一个怎样的人呢？"

瞿杰说："人们都以为我的女朋友一定是有倾国倾城之貌的淑女，不敢说一定门当户对，起码也是小家碧玉……可我就是让大家大跌眼镜，我的女朋友条件很差，长得丑，皮肤黑，个子矮，家里也很穷，但很有个性……得知我和她交朋友，家里非常反对。我说，我就是喜欢她，如果你们不认这个媳妇，我就不认你们。话说到这个份上，家里也只好默许了。总之，所有的人都不看好我的选择，但我义无反顾地爱她。可是，没想到，她却在11天前对我说，她不爱我了，她爱上了另外一个人……我以前听说过天塌地陷这个词，觉得太夸张了，就算地震可以让土地裂缝，天是绝对不会塌下来的，但是在那一瞬，我真正明白了什么是乾坤颠倒、地动山摇。我被一个这样丑陋的女人抛弃了，她找到的另外一个男人和我相比，简直就是一堆垃圾，不不，说垃圾都是抬举了他，完全是臭狗屎！"

瞿杰义愤填膺，脸上写满了不屑和鄙夷，还有深深的沮丧和绝望。

事情总算搞清楚了，瞿杰其实是被这种比较打垮了。我说："这件事的意义对于你来说，并不仅仅是失恋，更是一种失败和耻辱。"

瞿杰大叫起来："你说得对，就像八国联军入侵，我没放一枪一炮就一败涂地丧权辱国。如果说我被一个绝色美女抛弃了，我不会这么懊丧。如果说我被一个高干的女儿或是富商家的小姐甩了，我也不会这么愤慨。或者说啦，如果她看上的是一个美男帅哥大款爵爷什么的，我也能咽下这口气，再不干脆嫁了个离休军长，我也认了……可您不知道那个男生有多么差，我就想不通我为什么会败在这样一个人渣手里，我冤枉啊……"

"So it is a case of everything ready and waiting?"

"Exactly! Ready and waiting for the right girl to show up!"

"What kind of girl is your girlfriend?"

"Everybody had expected my girlfriend to be a startling beauty. And that if her folks were not on a par with mine, at least she would be the darling of a respectable small town household. As it turned out however, my girlfriend was a complete shock—small-sized, dark-skinned, ugly, and a poverty-ridden background to boot! But she's got character. My family was against the relationship from the start. But I stated that I loved her and that if they do not accept her, I will reject them and stick to her! So my family caved in. Who would have thought that eleven days ago, she stated that she had stopped loving me! That she was in love with someone else! I have always felt it an exaggeration when people say 'the sky collapsed and the earth caved in.' But at that moment, it was exactly how I felt. I was being rejected by a woman so much beneath me. What's more, the man for whom she had betrayed me was downright trash." As he spoke, Zhai Jie's face was a picture of contempt, but still with an undertone of dejection and despair.

Finally I have gotten to the bottom of the problem—Zhai Jie has been beaten by the comparison. I ventured: "As far as you are concerned, the impact of this affair is not really about losing your girlfriend. It is about the disgrace of losing such a girl to such a rival."

Zhai Jie shouted: "You got it! It was like being beaten without firing a shot. Like what the Eight Allied Armies had done to our country years ago. I could have taken it better if I had been jilted by a dazzling beauty—or the pampered heiress of a big boss. Or if she had ran away with a big shot. Even if she had secured an army officer with a fat pension, I could have swallowed it. But to be jilted for the sake of such a downright piece of trash! It is so unfair!"

看到瞿杰把心里话都一股脑地倾倒出来，我觉得这是很好的进展。我说：“我能体会到你深入骨髓的创伤，其实你最想不通的还不是失恋，是在这样的比较中你一败涂地、溃不成军！”

瞿杰愣了一下，说：“你的意思是说我的痛苦不是失恋引起来的？”

我说：“表面上看起来，是失恋让你痛不欲生。但是刚才你说了，如果你的前女友找的是一个条件比你好的男生，你不会这么难过。或者说，如果你的前女友自身的条件要是更好一些，你也不会这样伤心。所以，我要说，你的失败感和失恋有关，但更和其他一些因素有关。”

瞿杰若有所思道：“你这样一讲，好像也有一点道理。但是，如果没有失恋，这一切都不会发生啊。”

我说：“如果没有失恋，也许不会这样集中地爆发出来，但是恕我直言，你是不是经常在和别人的比较当中过日子？”

瞿杰说：“那当然了。如果没有比较，你怎么能知道自己的价值？”

我说：“瞿杰，这可能就是问题的关键所在了。其实，一个人的价值并不在和别人的比较之中，而是在自己的掌握之中。就拿你自己来当例子，你的条件和11天以前的你，有什么大的变化吗？”

瞿杰说：“除了睡不好觉，体重减轻头发掉了一些之外，似乎并没有其他的变化。”

我说：“对啊，那么，你对自己的评价有什么变化吗？”

瞿杰说：“当然有了。比如我觉得自己不出色、不优秀、不招人喜爱、前途黯淡了……”

我说：“你的篮球还打得那样好吗？”

It was a good sign. Now that Zhai Jie had let everything tumble out pell mell, we were finally getting somewhere. I said: "I understand your pain. What hurts most is not losing the girl—it is the fact of losing her to such a good-for-nothing!" Zhai Jie seemed struck by my remark and after taking it in, he ventured: "So you think I am not suffering from disappointed love?"

"On the surface, you seem to be dying for your lost love. But you have just said that if your ex-girlfriend had found someone better, it would have made sense to you. Or if her circumstances had been more advantageous, you would understand. So I would say that your feelings of defeat have something to do with losing your girlfriend, but it is not the whole story—there are other factors involved."

Zhai Jie stopped to think for a while and said: "You seem to have a point. But still, if my girl had not left me, all this would not have happened."

I conceded that if not for being jilted, there would not be such a crisis. "But still," I asked, "Are you in the habit of making comparisons between yourself and your peers? "

"Certainly," he replied, "Or else, how can you tell your own value?"

"This may be the crux of your problem! Actually a person's value does not depend on comparison with others. It is in one's own grasp. Take yourself for example. Is there much difference between what you are now from the person that you had been eleven days ago?"

"Well," said Zhai Jie, "apart from weight loss, sleep deprivation and some loss of hair, there's not much change in me..."

"Good," I went on, "and what about your self-esteem?"

"Of course I have loss of self-esteem! I feel that I am not outstanding among my peers. I am not popular. My future is pretty dim."

"Are you still good at basketball?"

瞿杰不解地说："当然啦。只是我这几天没有打篮球，如果打，一定还是那样好。"

我又说："你的歌唱得还好吗？"

瞿杰说："这个没有问题。只是我现在没有心思唱歌。如果唱哀伤的歌儿，也许比以前唱得还好呢。"

我接着说："你的学习成绩怎样呢？"

瞿杰好像明白了一些，说："还是很好啊。"

我最后说："你的个头怎样呢？"

瞿杰难得地笑出声来，说："您可真逗，就算我几天几夜不吃饭不睡觉，分量上减轻点，骨头也不会抽啊。"

我趁热打铁说："对呀，你还是那个你，只是这其中发生了失恋，一个女生做出了她自己的选择……我们还不完全知道她是因为什么做出这样的决定，但你只有接受和尊重这个决定，这是她的自由。两个相爱的人因为种种原因不能走到一起，固然是一个令人伤感的事情，但感情的事是不能勉强的。世上无数的人经受过失恋，但从此一蹶不振跌倒了就爬不起来的人毕竟有限。瞿杰，我看你面对的并不是担心自己以后找不到女朋友，而是更深的忧虑。"

瞿杰说："您说得太对了。寝室的男友知道我失恋的事，总是说，依你的条件这样好，还怕找不到好姑娘吗？别这么失魂落魄的，看哥们下午就给你介绍一个漂亮MM。他们不知道我心里的苦，并不是担心自己以后找不到老婆，而是想不通为什么会被人行使了否决权，我觉得自己在人格上输光了血本。"

"Of course I am! It's just that I have been out of practice these last couple of days."

"What about your voice?"

"No problem at all. The problem is that nowadays I am not in the mood. I might perform better if I sing some dirges."

"What about your grades?"

Zhai Jie seemed to have got the point. He stated: "As good as ever!"

"What about your height?" I ended up by asking.

Zhai Jie laughed out loud, "Even if I stop eating or sleeping for a couple of days, I might lose some weight, but my flesh and bones will still be what they are."

Now was my moment, and I seized it, saying:"Exactly! You are still what you were, except that your love affair fell through. The girl had decided to make a new choice. We do not know the reason, but it is her right, and we respect her decision. Two people in love may decide, for reasons of their own, to end their relationship. Partings are sad, but matters of personal affection can never be solved through coercion. All over the world, people fell in and out of love, but few collapse never to get up again. As for you, Zhai Jie, I suspect that you are not worried about finding someone else, I suspect it is something deeper."

"You have hit the nail on the head. My roommates keep saying: `You are the last person to have problems finding a new girl! Don't act so dejected. I'll bring you one this afternoon!' They don't know what's gnawing at me inside. I am not worried about finding a wife. I just can't understand why I was turned down! It is a blow to my self-esteem!"

我说："瞿杰，谢谢你这样勇敢地剖析了自己的内心，失恋只不过是个导火索，它点燃的是你对自己评价的全面失守，你认为女友的离开是地狱之门，从此你的人生走向黑暗。你看到她的新男友，觉得自己连一个这样的人都不如，就灰心丧气全盘否定了自己。"

在长久的静默之后，瞿杰的脸上渐渐现出了光彩，他喃喃地说："其实我并没有失败？"

我说："失恋这件事也许已成定局，但是人生并不仅仅是爱情，还有很多重要的事情在等待着你。再说，就是在爱情方面，你也并不绝望，依然有得到纯美爱情的可能性啊。"

瞿杰深深地点点头，说："从此我不会再从别人的瞳孔中寻找对我的评价，我会直面失恋这件事情……"

瞿杰还是被姐姐扶着走出咨询中心的。他的眼睛因为极度的困倦已经睁不开，靠在姐姐肩头险些睡着。大约一个半小时之后，工作人员说瞿杰的姐姐电话找我。我以为瞿杰有了什么新情况，赶紧接过电话。

瞿杰的姐姐说："我带着瞿杰，现在还在出租汽车上。"

我说："你们家这么远啊？"

瞿姐姐说："车已经从我们家门口路过好几次了。"

我说："那你们为什么像大禹治水一样，路过家门而不入？"

I said: "Zhai Jie, thank you for being so brave in dissecting your deepest feelings. Losing your girlfriend was just the tip of the iceberg—what it released was your loss of self-esteem. The breakup with your girlfriend was like a gate to hell. When you saw what her new boyfriend was like and realized that he had actually supplanted you, it was a blow to your self-esteem."

After a long silence, Zhai Jie's face gradually turned radiant.

"So I am not defeated," he murmured.

"The relationship is over, but life is not just about falling in love. There are more important things in life. Besides, there are other chances of love around the corner."

Zhai Jie nodded, "From now on," he said, "I am not going to bother about how other people view me. I will face up to the fact of being jilted and that's that!"

Zhai Jie still weak, was supported by his sister into the taxi. He was so sleepy that he could barely keep his eyes open; he nearly fell asleep on his sister's shoulder.

More than an hour later, one of the staff said that Zhai Jie's sister was on the phone. I rushed to the phone, thinking that a new problem had come up.

Zhai Jie's sister said: "I am still in the taxi with Zhai Jie."

"Is your house so far away?"

"The car had passed our house several times already."

"Are you trying to outdo Emperor Dayu, who passed by his own gate three times without stopping?"

瞿姐姐说："瞿杰一坐上出租汽车马上就进入了深深的睡眠，睡得香极了，还说梦话，说：我不灰心，我不怕……睡得口水都流出来了，好像一个甜甜的婴儿。这些天他睡不着觉非常痛苦，看到他好不容易睡着了，我不敢打扰他，就让出租车一直在街上兜圈子，绕了一圈又一圈，车费都快两百块钱了。我怕一旦把他喊起来，又进入无法成眠的苦海。可他越睡越深沉，没有一点醒来的意思，我也不能一直让车拉着他在街上跑。我想问问您，如果把他喊醒下车回家，他会不会一醒过来就又睡不着觉了？我好害怕呀！"

我说："不必担心，你就喊醒他下车回家吧。如果他还睡不着觉，就请他再来。"

瞿杰再也没有来。

"Zhai Jie had fallen into a deep sleep the minute he got into the car. He keeps mumbling in his sleep: 'I am not despondent' 'I am not afraid' while dribbling at the mouth. Just like a baby. These last few days, he had lost so much sleep that I don't have the heart to wake him up. I let him sleep on. The taxi has been circling around non-stop, the meter is crawling up to 200 yuan. I am afraid to wake him, in case he relapses into insomnia again. He is actually falling deeper and deeper into sleep, with no signs of waking. But I can't let the taxi roam the streets forever. The question is, shall I wake him up and take him home? Supposing he slips back into insomnia, what then? I am so scared!"

I said: "Not to worry. Wake him up and take him home. If he is unable to sleep, let him come back."

Zhai Jie never came back.

失眠失恋，你究竟失去了什么

Comparisons are Odious

在生和死之间，是孤独的人生旅程。

保有一份真爱，就是照耀人生得以温暖的灯。

任何成瘾都是灾难

有个年轻人，名叫安澜，他说自己干什么都会成瘾。

我要详细了解情况，就说："请打个比方。"

他说："我上学的时候就对网络成瘾。那时候，我每天起码有五小时要趴在网上，网友遍布全世界。"

我插嘴道："全世界？真够广泛的。"

安澜说："是啊。人们都说上网对学习有影响，可那时我的英文水平突飞猛进，因为要和国外的网友聊天，你要是英文不利索，人家就不理你了。"

我说："一天五小时，你还是学生，要保证正常的上课，哪里来的这么多时间啊？"

安澜说："很简单，压缩睡眠，我每天只睡五小时。我有单独的房间，电脑就在床边。我每天做完作业后先睡下，四小时之后，准时就醒了，一骨碌爬起来就上网，神不知鬼不觉的，到了天快亮的时候，再睡一小时回笼觉。爸爸妈妈叫我起床的时候，我正睡得香甜。很长时间，家里人看我白天萎靡不振的，都以为是上学累的，殊不知我的睡眠是个包子，外面包的皮是睡觉，里面裹的馅就是上网。"

我说："青少年正是长身体的时候，你这样睡眠不足，是要出大问题的。"

Addiction means Disaster

During one of my counseling sessions, a young fellow named An Lan confided to me that for him, anything that he enjoys would turn into an obsession. I asked for an example. He told me that when at school, he had been addicted to going online—he would spend five hours every day at the computer, his online friends were spread throughout the world.

I could not help exclaiming: "Throughout the world! That's really something."

"So it is," An Lan went on: "People say that going online affects your school work, but in my case, my English shot up. I had to chat with friends throughout the world, and if your English is not snappy, the kids ignore you."

"Five hours online! You were a student at the time. Where did you find so much spare time?"

"Easy," said An Lan, "Cut sleep to five hours. I had a bedroom to myself, and my computer was next to my bed. I went to bed right after finishing my homework. I woke up promptly after four hours' sleep and went online, and nobody the wiser. By daybreak, I stopped for another hour or so for sleep. When my mom or dad came knocking to wake me up for school, I would still be in deep sleep. I would go through the day pale and out of sorts. My parents assumed that I was overworked at school. Actually my problem was lack of sleep. For me, sleeping became a pretense, a cloak to hide the time I spent online."

I muttered that he was still in growing pains and that this lack of sleep would cause serious problems.

安澜说："还真让您说对了。后来，我就得了肾炎。因为不能久坐，我只好缩减了上网的时间。我休了学，急性期过了以后，医生建议我开始缓和的室外活动，慢慢地增强体力，我就到郊外或是公园散步。

"一个人在外面闲逛，就是风景再美丽、空气再新鲜，也有腻的时候。我爸说，要不给你买个照相机吧，一边走一边拍照，就不觉得烦了。家里先是给我买了个数码的傻瓜相机。果然，照相让人觉得时间过得很快，一只狗正在撒尿，一只猫正在龇牙咧嘴地向另外一只猫挑衅，都成了我的摄影素材。白天照了相，晚上就在电脑上回放，自己又开心一回。很快，这种简陋的卡片机就不能满足我的欲望了。我开始让家里人给我买好的机子，买各式各样的镜头……把自己认为好的照片放大。城周围的景物拍烦了，就到更远的地方去，我又迷上了旅游。

"后来我爸说，我这是豪华型患病，花在照相和旅游上的钱，比吃药贵多了。不管怎么样，我的病渐渐地好了。但是错过了高考，我就上了一所职业学校，学市场营销。毕业以后，我进了一家玩具公司。玩具这个东西，利润是很大的，只要你营销搞得好，拿比例提成，收入很可观。这时候，因为时间有限，到远处旅游和照相，变得难以实现，我就迷上了请客吃饭……"

我虽然知道咨询师在这时应该保持足够的耐心倾听，还是不由自主地小声重复："迷上了请客吃饭？"

说句实话，我见过各种上瘾的症状，要说请客吃饭上瘾，还真是第一次碰上。

"It was just as you said," exclaimed An Lan. "Soon, I was diagnosed with nephritis and could not remain sitting up for long. I had to cut the time spent online. Then I had to quit school. When the first acute attack was over, the doctor advised me to spend time outdoors to recover my strength. I would take walks in the park or the outskirts of town. No matter how fresh the air, or how beautiful the scenery, I soon became bored of being alone all the time. So my dad bought me a little digital camera. Taking random snapshots as I walked about got me out of boredom. A doggie peeing, a cat showing its teeth at a rival —they all served as subjects for photography. Back home at night, I would play back the shots and get some entertainment out of them. But soon I was dissatisfied with that simple contraption. I got a good camera with different lenses and took shots of views in and out of town, and back home I would enlarge them for fun. I would go further and further out of town in search of better views, and this actually turned into tourism. My father muttered that mine was really an expensive hobby, as the costs of my photography and 'tourism' had far exceeded the medical expenses.

"Despite the complaints about expenses, I did have a good recovery. But by then I had missed the National Test for Higher Education, so I attended a private business school where I studied marketing. On graduation, I got a job at a toy store. I was paid according to percentage of sales and enjoyed a good income. Under the circumstances, photography and tourism as pastimes were out of the question, so I became addicted to throwing dinner parties."

I was aware that as his counselor, I should hear him out, but I could not help exclaiming: "Addicted to throwing dinner parties!" To be honest, it was the first time I had ever heard of such an addiction.

安澜说：“是啊。我喜欢请客时那种向别人发出邀请，别人受宠若惊的感觉。喜欢挑选餐馆，拿着点菜单一页页翻过时的那种运筹帷幄的感觉，好像点将台上的将军，尤其是喜欢最后结账时一掷千金、舍我其谁的豪爽感。”

我思忖着说：“你为这些感觉付出的代价一定很高昂。”

安澜垂头丧气地说：“谁说不是呢？去年年底，我拿到了七万块钱的提成奖励，结果还没过完春节，就都花完了，我可给北京的餐饮业做出了杰出的贡献。最近，我们又要发季度提成了，我真怕这笔钱到了我的手里，很快就烟消灰灭。而且，酒肉朋友们散去之后，我摸着空空的钱包，觉得非常孤单。可是下一次，我又会重蹈覆辙，不能自拔。我爸和我妈提议让我来看心理医生，说我这个人爱上什么都没节制，很可怕。将来要是谈上女朋友也这样上瘾，今天一个明天一个，就变成流氓了。我自己也挺苦恼的，一个人，要是总这样管不住自己，也干不成大事啊，您能告诉我一个好方法吗？”

我说：“安澜，我知道你现在很焦虑，好方法咱们来一起找找看。你能告诉我像上网啊，摄影啊，旅游啊，请客吃饭啊这些活动，带给你的最初的感觉是什么吗？”

安澜说：“当然是快乐啦！”

我说：“让咱们假设一下，如果在那个时候，来了位医生抽一点你的血，化验一下你的血液成分，你觉得结果会怎么样？”

安澜困惑地吐了一下舌头，说：“估计很疼吧？结果是怎样的，就不知道了。”

An Lan said, "Yes, I like the feeling of sending out invitations and seeing the delight and surprise on people's faces. I like the experience of choosing a restaurant and going through the menu page by page, deliberating over the choices, like a general picking his team. I especially like the grand finale of throwing down the cash in a magnanimous gesture, as if saying who else if not I?"

I demurred, saying: "I suppose you must have paid a hefty price for this kind of 'feeling good'?"

An Lan looked downcast. "Exactly. On the eve of New Year, I received a bonus of seventy thousand yuan, but it all went down the drain and Spring Festival was barely a month away. I could say that I have really made a contribution to the success of the catering business in town. And now, our bonus for the second season is about to be handed out. I am really afraid that the money will disappear in a flash. The fact is, once the convivial gathering is scattered and I am left with an empty pocket, I feel very dejected. But I can't help myself. The minute there is money in my pocket, the cycle is on again, beyond my control...

"My parents have advised me to seek counsel, saying that this merry-go-round of dinners must be checked. Suppose I start dating, they warned, I will be seeing one girl today and a new girl tomorrow, and I will be taken for a womanizer. I myself am unhappy about the situation. I will never be up to much if this trend goes on. Do you have any advice?"

I said, "An Lan, I realize you are worried. Let's put our heads together and think of something. Tell me, what were your feelings when you first plunged yourself into going online, and photography, and tourism, and entertainment?"

"I was happy, of course!"

"Now supposing that at the time, a doctor showed up and wanted to take your blood for testing, how would you feel?"

Caught unawares, An Lan stuck out his tongue. "I suppose it will hurt. I won't be able to predict the result, though."

我说："抽血有一点疼，不过很快就会过去。我以前当过很久的医生，对化验这方面有一点心得。当人们在快乐的时候，内分泌系统会有一种物质产生，叫作内啡肽。"

安澜很感兴趣，说："您告诉我是哪几个字。"

我在一张纸上写下了"内啡肽"几个字。

安澜仔细端详着，说："这个'啡'字，就是咖啡的'啡'吗？"

我说："正是。咖啡也有一定的兴奋作用。"

安澜说："您的意思是说，每当我进入那些让我上瘾的活动的时候，我身体里都会分泌出内啡肽吗？"

我说："安澜，你很聪明，的确是这样的。内啡肽让我们有一种不知疲劳、忘却忧愁、精神焕发的感觉。这在短期内当然是很令人振奋的，但长久下去，身体就会吃不消。这就是很多染上了网瘾的人，最后变成茶饭不思、精神萎靡不振、体重大减、面黄肌瘦的原因啊。而且，因为人上瘾时，对其他的事情不管不顾，考虑问题很不理性，就会出现严重的后果，这也就是你在请人吃完饭之后精神十分空虚的症结。有的人工作成瘾，就成了工作狂；有的人盗窃成瘾，就成了罪犯；有的人飞车成瘾，就成了飙车族；有的人权力成瘾，就成了独裁者……"

安澜说："这样看来，内啡肽是个很坏的东西了。"

I told him that getting one's blood taken will hurt a little but that the pain will be gone very quickly. I had been a doctor and know something about blood testing. When people are happy, there are secretions from their endocrinal glands, called endorphin.

An Lan seemed very interested, and wanted me to show it to him in writing, so I wrote down on a piece of paper the three Chinese ideograms for endorphin. An Lan scrutinized them and said: "The *phin* here sounds like *fei*, the last ideogram for the word coffee."

I said yes, and coffee has the effect of stimulation.

An Lan asked: "So every time I get worked up in my hobby, it means that my body secretes this kind of endorphin?"

"Yes, An Lan," I told him, "the caffeine in coffee also has a stimulating effect."

An Lan asked: "Are you saying say that every time I get wrapped up in those activities which could be addictive, it means that my body is secreting that stimulant?"

I replied: "An Lan, you are smart, quick to get to the point. endorphin has the short-term effect of stimulating us and making us forget fatigue and unhappiness. In the long term, you get hooked. All such addictions are debilitating, causing loss of appetite, weight loss, and general decline of health. People throw everything to the winds when they are hooked. They stop thinking rationally and that may give rise to serious consequences. That is what happens to you after your dinner parties, you feel emptied out. In some cases, people become workaholics; others become kleptomaniacs and end up criminals. Then there are others who become part of the maniac drivers crowd. Still others become maniacs for power, and end up dictators..."

An Lan said: "This means that endorphin is a very bad thing?"

我说："也不能这样一概而论。人体分泌出来的东西，都是有用的。比如当你跑马拉松的时候，只要冲过了身体那个拐点，因为体内开始有内啡肽的分泌，你就不觉得辛苦，反倒会有一种越跑越有劲的感觉。比如有的科学家埋头科学实验，为了整个人类的发展做出了卓越贡献，在那种非常艰难困苦的条件下能够坚持下来，他的内啡肽也功不可没啊！"

安澜说："听您这样一讲，我反倒有点糊涂了。"

我说："任何事情都要有节制。比如，温暖的火苗在严冬是个好东西，可要是把你放到火上烤，结果就很不妙。如果你不想变成烤羊肉串，就得赶快躲开。再有，在干燥的沙漠里，泉水是个好东西，但要是发了洪水，让人面临灭顶之灾，那就成了祸害。对于身体的内分泌激素，我们也要学会驾驭。这说起来很难，其实，我们一直在经受这种训练。比如你肚子饿了，经过一个烧饼摊，虽然烤得焦黄的烧饼让你垂涎欲滴，但是如果你没买下烧饼，你就不能抢上一个烧饼下肚。如果你看到一个美丽的姑娘，虽然你的性激素开始分泌，你也不能上去就拥抱人家。所以，学会控制自己的内啡肽，也是成长的必修课之一啊。"

I said: "We shouldn't make these generalizations. Whatever the human body secretes, it has its uses. For instance, while running the Marathon, your body will be secreting endorphin after you have passed that critical turning point, and then you will stop feeling so tired. Actually you may even feel lighter on your feet. It is the same with scientists. As they bury themselves in research on subjects which benefit mankind, it is also thanks to the effect of endorphin that they could persist in their work under difficult circumstances."

An Lan murmured that he was getting a bit confused.

I said : "There is a limit to everything. Fire is good for winter, right? But I bet you don't want to be held over the fire, unless you want to be baked. Again, water is a good thing in the desert, but you don't want to be caught in a deluge. It is the same with the hormones in our bodies—we must learn to control. It sounds difficult to achieve; actually we are constantly being trained unawares. Say you are hungry and you pass by a stall with tantalizingly crispy baked buns on display. No matter how hungry, you will not snatch one unless you have paid for it. When you see a beautiful girl, your hormones might start working, but you will not go up and give her a hug. Controlling your endorphin is part of growing up."

听到这里，安澜若有所思地拿起那张纸，看了又看，说："这个内啡肽的'啡'字和吗啡的'啡'字，也是同一个字。"

我说："安澜，你看得很细，说得也很正确。成瘾这件事，最可怕的是毒品成瘾。吗啡和内啡肽有着某种相似的结构，当有些人靠着毒品达到快乐巅峰的时候，他们就步入了一个深渊，这就更要提高警惕了。当然了，网瘾和毒品成瘾还是有一定的区别的。不过，一个人要身体健康和心理健康，对所有那些令我们成瘾的事物都要提高控制力，要有节制。"

那天告辞的时候，安澜说："我记住了，任何成瘾都是灾难。"

As I finished speaking, An Lan picked up the piece of paper, looked closely at the writing and said: "The *phin* in the term endorphin is the same ideogram as the *phine* in morphine."

I said, "An Lan, you are a sharp observer and you got it right. Talking of addictions, the worst addiction is drugs. The structure of morphine and that of endorphin have something in common. People who are addicted to drugs are actually in hell. That is something that we must beware of. Of course, addiction to going online is something different from reliance on drugs. But still, in order to maintain bodily and mental health, we should keep a distance from anything that is addictive. We must practice self-control—everything in moderation."

As he left, An Lan murmured to himself: "Keep this in mind: addiction means disaster."

温暖的荆棘

这一天，咨询者迟到了。我坐在咨询室里，久久地等候着。通常，如果来访者迟到太久，我就会取消该次咨询。因为是否守时，是否遵守制度，是否懂得尊重别人，都是咨询师需要以行动向来访者传达的信息。试想一下，如果一个人在没有不可抗力的情况下，对准备帮助自己的人都不能践约，你怎能期待他有良好的改变呢？再说，重诺守信也是现代社会的基本礼仪。因为等得太久，我半开玩笑地问负责安排时间的工作人员说："这是一位怎样的来访者，为什么迟到得这样凶？"

工作人员对我说："您不要生气，千万再等等他们吧。"

我说："他们是谁，好像打动了你？为什么你的语气充满了柔情，要替他们说好话？我记得你平常基本上是铁面无私的，如果谁迟到超过15分钟，你都会很不客气。"

工作人员笑着说："我平常是那么可怕吗？就算铁石心肠也会被那个小子感动的。他们是一双外省的青年男女，失恋了，一定要请你为他们做咨询，央求的时候男孩嘴巴可甜了。现在他们坐在火车上正往北京赶呢。大雨倾盆阻挡了列车的速度，小伙子不停地打电话道歉。"

我说："像失恋这样的问题，基本上不是一次两次咨询就可以见到成效的。他们身在外地，难以坚持正规的疗程，不知道你和他们说过吗。"

Tender Thistles

The appointees were late that evening. I had been sitting in my office waiting. Usually I would cancel the appointment if the caller does not show up on time.

Keeping appointments is a matter of mutual respect and compliance with the rules of the Center, as well as a test of a person's sense of timing—all messages that the counselor tries to instill into the people seeking counsel. If one does not even show up on time for getting help—how can such a person be expected to make a change for the better? After all, showing up on time is one of the basics of good behavior. I asked the staff member: "To keep us waiting like this, what kind of visitor could it be?"

The staff member begged me to be patient and wait a little longer. I asked who these people are who had obviously touched a chord in him—he being a no-nonsense character who would not even tolerate a 15-minute delay? He smiled and asked me if he was really such a bluebeard. He said that the latecomers, a boy and a girl, would melt a heart of stone. They had both been disappointed in love, he was told, and are on a train coming to me for guidance—they were delayed by rain and the young fellow had phoned several times to apologize.

I reminded him that problems such as disappointed love cannot be dealt with in a few days and that he should let those people know what they are in for, coming from out of town.

工作人员急忙说："我都讲了，那个男生叫柄南，说他们做好了准备，可以坚持每周一次从外地赶来北京。"

原来是这样。那就等吧。原本是下午的咨询，就这样移到了晚上。他们到达的时候，浑身淋得落汤鸡一般，女孩子穿着露肚脐的淡蓝短衫和裤腿上满是尖锐破口的牛仔裤，十分前卫和时髦的装束，此刻被雨水浇在身上，像一个衣衫褴褛的丐帮弟子。她叫阿淑。

柄南也被淋湿，但因他穿的是很正规的蓝色西裤和白色长袖衬衫，虽湿但风度犹存。柄南希望咨询马上开始，这样完成之后，还能趁着天不算太黑去找旅店。

工作人员请他们填表。柄南很快填完，问："可以开始了吗？"

我说："还要稍微等一下。有个小问题，吃饭了吗？"

"吃了。"两个人异口同声地回答。

我又问："吃的是哪一顿饭呢？"

他们回答说："中午饭。"

我说："现在已经过了吃晚饭的时间。空着肚子做咨询，你们又刚刚经历了这么大的风雨，怕支撑不了。这里有茶水、咖啡和小点心，先垫垫肚子再说。"

两个人推辞了一下，可能还是冷饿占了上风，就不客气地吃起来。点心有两种，一种有奶油夹心，另一种是素的。阿淑显然是爱吃富含奶油的食品，把前一种吃个不停。柄南只吃了一块奶油夹心饼之后就专吃素饼了。看得出，他是为了把奶油饼省给阿淑吃。其实点心的数量足够两个人吃的，他还是呵护有加。

等到两人吃饱喝足之后，我说："可以开始了。"

柄南对阿淑说："你快去吧。"

我说："不是你们一起咨询吗？"

My staff member said that he did warn them about the distance, but the young fellow was prepared for the problem—they were planning to take the train on a weekly basis. That being the situation, I had no choice but to wait.

They arrived by night, thoroughly drenched.

The girl, Ashu, was dressed in cowboy shorts poked with holes in the knee and a tattered light blue top, exposing her belly button.

The young man Bingnan, in formal dark blue jacket and white shirt, though drenched was cool and collected. He wanted the counseling to start immediately, so that they could go and find a hotel before it got too dark.

The young man quickly filled the form, and asked whether the counseling could start.

"Just a minute," I asked, "Have you eaten?"

"Yes," both of them answered.

"When?"

"Noon."

Since it was already past supper time, I suggested that they take a snack at the clinic, before the counseling began. After a little show of polite refusal, hunger and cold had the upper hand, and the two young people sat down and helped themselves. Ashu obviously enjoyed the cream-filled bakery, and Bingnan seeing her preference, saved them for her, taking the plain bakes himself.

We started when they had their fill of food and drink.

I asked: "Are you getting counsel together?"

柄南说："是她有问题，她失恋了。我并没有问题，我没有失恋。"

我说："你是她的什么人呢？"

柄南没有正面回答我的问题，只是说："她是我的女朋友。"

我说："难道失恋不是两个人的事吗？为什么她失恋了，你却没有失恋？"

柄南说："你慢慢就会知道的。"

我真叫这对年轻人闹糊涂了。好比有一对夫妻对你说他们离婚了，然后又说女的离婚了，男的并没有离婚……恨不能就地晕倒。

咨询室的门在我和阿淑的背后关闭了。在这之前，阿淑基本上是懈怠而木讷的，除了报出过自己的名字和吃了很多奶油饼外，她的嘴巴一直紧闭着。随着门扇的掩合，阿淑突然变得灵敏起来，她用山猫样的褐色眼珠迅速睃寻四周，好像一只小兽刚刚从月夜中醒来。在我面前坐定，伸直她修长的双腿之后的第一句话是："您的这间屋子隔音性怎么样？"

我还是第一次碰到来访者问这样的问题，就很肯定地回答她："隔音效果很好。"

阿淑还不放心，追问道："就是说，咱们这里说什么话，外面绝对听不到？"

我说："基本上是这样的，除非谁把耳朵贴在门扇上。但这大体是行不通的，工作人员不会允许。"

阿淑长出了一口气，说："这样我比较放心。"

我说："你千里迢迢地赶了来，有什么为难之事呢？"

阿淑说："我失恋了很想走出困境。"

我说："可是看起来你和柄南的关系还挺密切啊。"

Bingnan said: "She has a problem, not me. She is disappointed in love."

I then asked about their relationship.

Bingnan did not elucidate, mumbling that she was his girlfriend.

I asked: "How come she is disappointed in love and you are not involvcd? I assume that disappointment in love involves two people."

Bingnan said that I will understand if I hear her out

I was totally at sea. It sounded as if a couple had divorced and then insisted that the woman was divorced but not the man. Beats me.

The door closed behind Ashu and myself as we sat down for consultation.

So far, Ashu had been silent and apathetic. Apart from giving her name and eating a lot of creamy bakery, she had barely opened her mouth. But as soon as the door closed behind us, she became a different person. She looked around with her brownish cat-like eyes and asked: "Is this room soundproof?" It was actually the first time that someone had asked this question, and I assured Ashu that it was.

Ashu still wanted assurance, asking if the room was absolutely soundproof.

I re-assured her that it was—unless one of the staff put his ear to the door, something unheard of so far, and most unlikely.

Ashu gave a sigh of relief, and I asked what had sent her coming so far for help.

Ashu said that she had been disappointed in love, and was seeking to get out of depression.

"But to me your relationship with Bingnan seems quite close..."

阿淑说："我并不是和他失恋了，是和别人。那个男生甩了我，对此我痛不欲生。柄南是我以前的男友，我们早已不来往好几年了。现在听说我失恋了，就又来帮我，陪着我游山玩水，看进口大片，吃美国冰淇淋，您知道这在外省的小地方是很感动人的。包括到北京来见您，都是他的主意……"阿淑说话的时候不时地看看门的方向，好像怕柄南突然把门推开。"

我说："阿淑，谢谢你对我的信任，我对你们的关系比较清楚一点儿了。那么，我还想更明确地听你说一说，你现在最感到困惑的是什么呢？"

阿淑说："天下没有免费的午餐，当然也没有免费的人陪着你走过失恋。现在的问题是我要甩开柄南。"

说到最后这一句话的时候，阿淑把声音压得很低，凑到我的耳朵前，仿佛我们是秘密接头的敌后武工队员。

我在心底忍不住笑了——在自己的咨询室里，我还从来没有过这样鬼鬼祟祟的样子呢。面容上当然是克制的，来访者正在焦虑之中，我怎能露出笑意？我说："看来你很怕柄南听到这些话？"

阿淑说："那是当然了。他一直以为我会浪子回头和他重修旧好，其实，这是根本不可能的。谢谢他，我已经从旧日的伤痕中修复了，可以去争取新的爱情了，但这份爱情和柄南无关。我到您这儿来，就是想请您帮我告诉他，说我并不爱他。我是失恋了，但这并不等于他盼来了机会。我会有新的男朋友，但绝不会是他。"

我看她去意坚决，就说："你已经想得很清楚了？"

阿淑说："是的，很清楚了，就像白天和黑夜的分割那样清楚。"

"It is not Bingnan," she assured me. "It is someone else. He dumped me for another girl. I just cannot live with the pain. Bingnan was my previous boyfriend; we had stopped seeing each other for the last several years. He heard about my being dumped and came to my help. He took me on tours, treated me to American ice cream, and we watched foreign movies. Such treats are a big deal in the small town where I live. Even coming to consult you was his idea."

Ashu kept checking the door as she spoke, as if it could fly open without notice

I said: "Thank you Ashu, for trusting me. Now I am beginning to get the whole picture. But tell me in your own words, what is your most urgent problem right now?"

"There is no such thing as a free lunch," said Ashu, "and you do not expect anyone to help you get over a disappointment in love—for free! *My most pressing problem now is to get rid of Bingnan*." Ashu lowered her voice and whispered the words into my ears, as if we were underground agents passing information.

I couldn't help laughing up my sleeve—to be acting in such a furtive way! And in my own office too! But I had to control myself. My visitor was in an agony of anxiety—how could I show amusement at her expense?

I said to Ashu: "You don't want Bingnan to know!"

"Certainly not! He expects me to go back to him. But it is impossible. I am thankful to him. I myself have gotten over this recent break-up. I am actually here to ask you to pass on this information—to tell Bingnan that I do not love him, that being disappointed in love does not mean that I am turning to him. I will find my true love eventually, but it will not be Bingnan."

I asked her: "So you have thought it through?"

"Yes," she answered "My attitude and his expectations, they are as far apart as night and day!"

我说："这个比方打得很好，让我明白了你的选择。但是，我还有一点很疑惑，你既然想得这样清楚，为什么不能说得同样清楚呢？你为什么不自己对柄南大声说分手？你们朝夕相处，肯定不止有100次讲这话的机会。为什么一定要千里迢迢地跑到北京来，求我来说呢？"

阿淑把菱角一样好看的嘴巴撇成一个外八字，说："您怎么连这都不明白？我不是怕伤害他嘛！"

我说："你很清楚你不承认是他的女朋友，就伤害了柄南？"

阿淑说："几年前，我第一次离开他时，他几乎吞药自杀，好不容易才缓过神来。这一次，真要出了人命关天的事，我就太不安了。"

我说："阿淑，看来你的内心深处，还是一个善良的女孩。只是，当你深陷在失恋的痛苦中的时候，你明知自己无法成为柄南的女友，还是要领受他的关爱和照顾，因为你需要一根救命的稻草。现在，你浮出了旋涡，就想赶快走出这种暧昧的关系。只是，你不愿意看到这种悲怆的结局，你希望有一个人能代替你宣布这个残忍的结论，所以你找到了我……"

阿淑说："您真是善解人意。现在，只有您能帮助我了。"

我说："阿淑，真正能帮助你的人，只有你自己。虽然我非常感谢你的信任，但是，我不能代替你说这样的话，你只有自己说。当然了，这个'说'，就是泛指表达的意思。你可以选择具体的方式和时间，但没有人能够替代你。"

阿淑沉默了半天，好像被这即将到来的情景震慑住了。

她吞吞吐吐地说："就算我知道了这样做是对的，我还是不敢。"

"You have made it clear. But my next question is, since you have thought things through, why don't you go ahead and tell him in your own words? You have been together these last few days and had plenty of opportunities to speak. Why come all this way to Beijing to make me talk to him?"

Ashu pursed her pretty lips: "How come you still don't understand! Can't you see that I am afraid of hurting him?"

"Are you sure that he will be hurt?"

"A few years ago, when I broke up with him, he almost committed suicide! He ultimately recovered. If something like that happens again, I will never forgive myself."

I said: "Ashu, I can see that you are good at heart. But when you were suffering from disappointed love, you turned to Bingnan. You accepted his care without any intention of loving him back. At the time, you grasped at any straw to save yourself from total collapse. Now that you have come through, you want to shake off this ambiguous relationship. But you do not want to face the music—you want me to do the dirty work of announcing the end of the relationship..."

Ashu acknowledged that I was right and that only I can help her.

I said: "Ashu, thank you for your trust. But the only person who can help you is you. No one can spell out the words for you. Others might suggest the right time and place, but no one can do it for you."

Ashu kept silent, as if stunned by the looming prospect. She muttered, "Even if I know it is the right thing to do, I am still afraid..."

我说："阿淑，咱们换一个角度想这件事。如果柄南不愿意和你保持恋人的关系了，你会怎样？"

阿淑说："这是不可能的。"

我说："世上万事皆有可能。我们现在就来设想一下吧。"

阿淑思忖了半天，说："如果柄南不愿意和我交朋友了，我希望他能当面亲口告诉我这件事。"

我说："对啊，己所不欲，勿施于人。如果柄南找到一个第三者，托他来转达，你以为如何呢？"

阿淑咬牙切齿地说："那我会把第三者推开，大叫着好汉做事好汉当，千方百计找到柄南，揪住他的衣领，要他当面锣对面鼓地给我一个说法，一个解释，一个理由，一个结论！"

我说："谢谢你的坦诚，答案出来了。失恋这件事，对于真心投入的男女来说，的确非常痛苦。但再痛苦的事件，我们都要有勇气来面对，因为这就是真实而丰富多彩的人生的本来面目。困境时刻，恋情可以不再，但真诚依旧有效。对于你刚才所说的'四个一'，我基本上是同意一半，保留一半。"

阿淑很好奇，说："哪一半同意呢？"

我说："我同意你所说的——对失恋要有一个结论、一个说法。因为失恋这个词，你想想就会明白，它其中包括了个'失'字，本质就是一种丧失，有物质更有精神的一去不复返，有生理更有心理的分道扬镳。对于生命中重要事件的沉没，你需要把它结尾。就像赛完了一项马拉松或是吃完了一顿宴席，你要掐停行进中的秒表，你要收拾残羹剩饭、刷锅洗碗。你不能无限制地孤独地跑下去，那样会把你累死。你也不能面对着曲终人散的空桌子发呆，那渐渐腐败的气味儿会像停尸间把人熏倒……

阿淑说："这一半我明白了，另一半呢？"

I said: “Ashu, let us take another angle. Supposing Bingnan broke up his friendship with you...”

She said: “That is impossible!”

“Everything is possible. Let us imagine the situation.”

Ashu thought for a while and said: “If Bingnan stops being my friend, I want him to tell me so in his own words!”

“Exactly so! Do not do to others what you do not want for yourself! Supposing when you were dating Bingnan, he found someone else and asked a third party to let you know. How would you feel?”

“I would push aside the third party, and demand that Bingnan himself give me an explanation, an excuse, a reason, a wrap-up!”

“Thank you for your frankness. Now we have an answer. To someone who had been deeply in love, breaking up is very painful. But however painful the break-up, one must face the fact if one were ever caught in such a situation. It is part of the multi-faceted face of life itself. In certain circumstances, love may vanish, but sincerity must never fail. As to the four indispensables that you have mentioned, I would accept two of them and reject the other two.

Ashu was curious as to which of the ‘indispensables’ were retained and which rejected.

“I agree with you that one must have an explanation and a wrap-up, for falling out of love. Note the word ‘out’ in the phrase ‘falling out of love.’ It denotes loss, both material and mental, both physical and emotional. You need a finishing line, as in a marathon, or as in a banquet, both of which must have an ending. One does not keep running on and on, just as one does not keep eating non-stop. Otherwise one would either die of exhaustion or keep sitting at table until the food turns bad.”

“Now I understand the part about the necessity for a ‘wrap-up’ and an ‘explanation,’ but what about the other half of the demand—an excuse and a reason?” asked Ashu.

我说：“我持保留意见的那一半，是你说在失恋分手的时候要有一个解释一个理由。”

阿淑说：“我刚才还说少了，一个解释一个理由哪里够用？最少要有十个解释十个理由！轰轰烈烈的一场生死相依，到头来悄无声息地烟消云散了，还不许问为什么，真想不通！郁闷啊郁闷！”

我说：“我的意思不是瞒天过海什么都不说，不是让大家蒙沉云里雾里，死也是个糊涂鬼。人心是好奇的，人们都愿意寻根问底，踏破铁鞋地寻找真谛。这在自然科学方面是个优良习惯，值得发扬光大，但在情感问题上，盘根问底要适可而止。失恋分手已成定局，理由和解释就不再重要。无论是性格不合还是家长阻挠，无论是两地分居还是第三者插足，其实在真正的爱情面前，都不堪一击。没有任何理由能粉碎真正的伴侣，只有心灵的离散才是所有症结的所在。理由在这里不再重要，重要的是你要接受现实。”

阿淑点点头说：“我明白您的意思了。我应该有勇气面对自己的失恋，我不能靠着柄南的体温来暖和自己。况且，这体温也不是白给的，他需要我用体温去回报。温暖就变成了荆棘。”

我说：“谢谢你这样深入地剖析了自己，勇气可嘉。特别是体温这个词，让我也百感交集。本来你们重新聚拢在一起，是为了帮你渡难关，现在，一个新的难关又摆在你们面前了。”

阿淑身上的湿衣已经被她年轻的肌体烤干了，显出亮丽的色彩。她说：“是啊，我很感谢柄南伸出手来，虽然这个援助并不是无偿的。现在，我要勇敢地面对这件事了，逃避只会让局面更复杂。”

"I do have reservations about the demand for a 'reason' and an 'excuse' for lovers' parting."

"On the contrary," said Ashu still smarting from a love affair gone sour, "I actually feel that one reason and one excuse are far from enough! I would demand at least ten!" exclaimed Ashu, "How can an earth-shaking love affair like mine dissolve into thin air without so much as a ripple? And not even allowed to ask for a reason! It is hard to swallow!"

"I did not mean that we should keep mum about the whole thing and leave everybody in the dark. People are curious by nature. They like to get to the bottom of things. This attitude is to be encouraged in the sciences. In affairs of the heart, however, discretion is advisable. Once the relationship is over, reasons and explanations become superfluous. Be it difference in temperament or parental opposition, be it long separation or a new attraction—nothing will shake true love. But once the heart is changed, however, nothing can retrieve lost love. Reasons are beside the point. One must accept that it is over."

Ashu nodded. "Now I am beginning to understand. I must pick up my courage and deal with being jilted. I should not use Bingnan's warmth for consolation. Besides, the consolation is not free. Bingnan expects a return. And then the consolation becomes a pain."

I said: "It is brave of you to analyze yourself so thoroughly, especially your perception of the nature of Bingnan's consideration. Your getting together was supposed to help you recover from your disappointment, but it has raised new complications."

By then Ashu's outfit had been dried by her own body heat. "Yes," she said, "I am grateful to Bingnan for showing up when I needed support. But the support is not free. I am recovered from my disappointment. Now I must deal with Bingnan; there is no running away from the problem."

我说："好啊，祝贺你迈出了第一步。天色已经不早了，你们奔波了一天，也需安歇。今天就到这里吧，下个星期咱们再见。"

阿淑说："临走之前，我要向您交一个功课。"

这回轮到我摸不着头脑，我说："我并不曾留下什么功课啊？"

阿淑拿起那张登记表，说："这都是柄南代我填的，好像我是一个连小学二年级都没毕业的睁眼瞎，或是已经丧失了文字上的自理能力的废人。他大包大揽，我看着好笑，也替他累得慌。可是，我不想自己动手。我要做出小鸟依人的样子，让柄南觉得自己是强大的，让他感觉我们的事情还有希望。现在，我知道在这个问题上，我利用了柄南，自己又不敢面对，就装聋作哑得过且过。现在，我自己来填写这张表，我不需要您代替我对他说什么了，也不需要他代理我填写什么了。"

我真是由衷地为阿淑高兴，她的脚步比我最乐观的估量还要超前。

看着他们的身影隐没在窗外的黑暗中，我不知道他们还会并肩走多远，也不知道他们的道路还有多长，但我想他们会有一个担当和面对。工作人员对我说："你倒是记着让来访者吃点心当晚饭，可是，你自己到现在什么也没吃啊。"

我说："工作之前不会觉得饿，工作之中根本不会想到饿。现在工作已经告一段落，饿和不饿也不重要了。"

I agreed, saying: "Good for you, you are on the right path. It is getting dark and you have had a long day. Supposing we stop here, and meet again next week."

"Before leaving, I must hand in a piece of homework," said Ashu.

I was mystified, murmuring: "I never left you any homework…"

Ashu picked up the form, which had been filled by Bingnan on her behalf, "He takes care of everything as if I am a child, or illiterate, or totally helpless." she said, "I play the helpless damsel and let him do it, so that he feels good about himself, even thinking that we might get together again. I am afraid to disillusion him on this count. I realized that I have used him, and that I was too cowardly to face the fact. Now I am going to fill a form on my own. I don't need you to say anything to him on my behalf, and I don't need him to fill forms for me."

I was glad on Ashu's behalf. She has moved forward faster than I expected.

I saw the moving shadows of the two of them outside my window. I have no idea how long they will be walking together, nor how far they will go. But I was sure that they will face up to the facts.

My staff interrupted my thoughts saying: "You have offered them pastry, but you yourself have not taken a bite yet."

I said: "You know that I don't eat before work. I never think of food during work. And now that the work is over, it doesn't matter whether I eat or not!"

温暖的荆棘

Tender Thistles

我们对自身的认识，也是抽丝剥茧，

需要水落石出的过程。

剩女出嫁

现如今的剩女，多数是上个世纪70年代的产品，她们是时代列车的特殊乘客，前无古人，后无来者。容我大胆地预测一下，以后的年代，难得有这样大规模的剩女集团军了。因为这一代剩女，已经用自己的蹉跎经历，提示了未来整整100年内的中国女子，等待就是放弃，不能坐失良机。不信，你看看如今的征婚启事，刚刚20岁出头的小闺女，就在那里张罗着虚空的缘分和脚踏实地的有房有车了。

如果是一筐苹果，任人挑选，那么剩下的一定是皱烂有虫的酸果。如果是一座楼宇，尾房多半是朝向和风水上有着这样那样的纰漏。如果真是一点褶子都没有“尾”在那里，买家几乎要怀疑那是否是凶宅，反倒越发地不放心。总之，剩的就是坏的，几乎是铁律。

但是，剩女可不是这样，令人扼腕。她们往往是优秀而雅致的女子，学识高，钱多，容颜身材也多居上乘。

这就令人不解。众人不解，剩女们也不解。我们这样出类拔萃，为什么反倒嫁不出去？男人啊，你们真是集体雀盲了吗？

Marrying off the "left-over" Women

Nowadays, the so-called "left-over" women are mostly relics of the seventies. They are a unique group of passengers in the journey of Life. And, if I may say so, the coming years will never see a repeat of such a phenomenon—a generation of young women left unmarried, and on such a massive scale.

The phenomenon will not be repeated because, based on their own frustrations, this current group of "left-over" women had left a message for their sisters of the new century, warning them that one must seize the moment, and that waiting is tantamount to self-defeat.

Indeed, their warning is being heeded. Just look at the ads for partners, put up by young women in their early twenties—a lot of gabble about affinity in a previous life and so on, but keeping a cold calculating eye on the basics—a house and a car.

Coming back to the issue of the "left-overs", now if it is a basket of apples for free, it goes without saying that the ones left untouched are rotten. Likewise, if it is a question of housing, the apartments left unsold are probably drafty or cursed with bad *fenshui*.

In this case however, the "left-over" young women are usually highly educated, well bred, well-off with handsome looks and good figures—and yet unmarriageable.

This is inexplicable to the general public, as well as a mystery to the young women themselves—"We are a cut above others in every way, why can't we get married? Oh, men, are you collectively blind?"

有人说，剩女不是时代或是他人造成的，剩女是她们自己造出来的。

剩女这传统，古而有之。

那些最美丽的女子，有些原本是悄然等着进宫的。她们怀揣着幽深梦想，期待有一天会守候在君王之侧，倘生下龙子，就有可能贵为国母。那些稍逊一等的女子，等着被达官贵人或是公子王爷看中，步入深宅大院，或是红袖添香，或是作威作福。特别爱洁净清心寡欲的女子，因了种种的未如意，多半入了尼姑庵，后来又有了当修女这扇小门，进入后从此离人间远、离天堂近了。靠出卖体力帮工过活的女子，就死心塌地当一辈子的管家或是女佣，也马马虎虎落得个轻手利脚温饱尚可。

现如今，这些个秀美女子的太平门，几乎都被堵死了。过去上等的姑娘是可以等的，好容貌好学识好脾气，如同香花一样在十里八乡传布，于是就有闻讯而来的媒人，首先为乡绅或是官吏等衣食丰饶之家来说亲。那时是可以三妻四妾的，那时的女人都活得短暂，因了生育或是疾病，往往早早就谢世了，于是候补上来的夫人，也有可能铺排出山花烂漫的日子。纵是自己有几腔苦水想倾倒，深宅大院的，也流淌不出来，无人知晓，传播出去的还是好福气的名声。

There are people who claim that the problem of these women should not be attributed to others, and that they had made their own beds and must lie therein.

Actually we Chinese have an age-old tradition of women staying unmarried. In the old days, there used to be beautiful young girls who do not rush to wed, nursing a hope of being selected to serve his Royal Majesty. If blessed with male offspring, such a girl might end up as Mother of the Nation. Girls on a lower tier of attractions could catch the eye of a member of the nobility and end up either in a secret love bower, or outright mistress of a noble household. Then there was another type of young women, placid and passionless, who take refuge in a Buddhist temple when Fate is against them. In later years, they had the additional choice of a Catholic convent, where the girl will find herself close to Heaven and far from the *madding crowd* on Earth. Then there had been women of lowly birth who decide to support themselves through a life of domestic service, enjoying a reputation for being light-footed with nimble fingers.

Nowadays, of course, those exits for women are closed.

In the old days, girls with good looks, good temper and good education could afford to wait. In their own neighborhood, their reputation spread like the scent of flowers and matchmakers would come knocking at the door, making offers on behalf of members of the local gentry. In those times, multiple wives and concubines were legal. Moreover, lifespan was short for women, prone as they were to early death through childbirth or untreated illness. Thus space was open for second wives, who behind enclosed courtyards, could carve out a life of comfort for themselves —or drag out the days in silent misery, and no one the wiser.

如今的剩女们，还在苦心经营地等，健身啊练瑜伽啊学插花食道兼修肚皮舞啊……却忘了一个冷酷现实——当你把自身修炼得珠圆玉润时分，那些最优等的男子，已被人捷足先登追求走了。记得哪位伟人说过，男子长期保持未婚状态，其实是有不忠和通奸的嫌疑。真正的好男子，极少长久孤身。不仅事业上打拼超越，他们也是有道德和良知的。他们有好的体力，自然也有好的情欲，为什么不顺应天时呢？结了婚，过得自在，也不会轻易抛弃了结发的妻，娶回一个踌躇意满的剩女为孩儿的后娘。

偏偏现代的医学技术又突飞猛进地发展，人的平均寿命已经超过了70岁，特别是城市的女子，干脆瞠目结舌地过了80这道大坎，这就让剩女走续弦预备队的可能性也大大地渺茫了。

剩女们初长成的时光，正是中国开放和发展风起云涌的时代。剩女们那一刻风华正茂，青翠欲滴。她们笃信时代不同了，男女都一样。她们认为只要自己有学识有力量，就可以逢山修路遇水架桥，有本领就有了一切。却不想，古老的法则在择偶问题上那样僵化，朴朴素素仅一条——先下手为强，后下手遭殃。除了捷足先登，再无良方。潜伏很深的总裁和博士，那时还是潜伏很深的青涩小伙儿，峥嵘之角只是额头柔软的小凸起，一碰就破。

The "left-over" women of today, however, are still preparing themselves for things to happen, spending time on health clubs, yoga lessons, floral arrangements, belly dancing and what not. By the time they have fixed themselves up to perfection, all the best and brightest of the eligible men have been snatched up. Some Sage had said that men who remain single are probably tangled in affairs. He backed his statement by arguing that a good man rarely remains single for long, since he not only has moral sense and excels at his job, but also stays healthy with a vigorous sexual appetite, and is likely to marry young. The Sage further states that the good man is unlikely to betray his wife, or marry a "left-over" woman as stepmother to his motherless child.

Thus we are left with the phenomenon of a multitude of single women. Unfortunately for these young women, with the progress of science people have a longer life span, especially in urban areas, where lifespan could easily stretch into the seventies, or even eighties. Thus the chance of being second wife to a desirable widower is also reduced.

The current generation of "left-over" women had first bloomed in the eighties when China had been committed to development and opening up to the world. They were young. They believed that times had truly changed, and that men and women were equal. They assumed that so long as they had the qualifications, they could move mountains and nothing could stop their drive to achieve. Little did they know that as far as marriage was concerned, the same age-old tenet held true: "The early bird gets the worm." At the time, the big boss and the academic genius of today were still in their prime, but by now have long been plucked and appropriated.

我的一位老朋友之女，说想向我咨询一下自己缘何成了剩女。我说：“在咨询室之外，我不接受咨询的任务，尤其不接受熟人的咨询。那样，对我不好，对你更不好。”

她叫青桦，说：“对您不好，我可以明白。比如打扰您的休息啊，您说深说浅都不是啊，还有，您不好意思拿咨询费了。”

我说：“青桦你说得都不错，但最主要的是，这违背了咨询的法则。咨询师和来访者，不能有双重关系。你到底是我的咨客呢？还是我的熟人？这个问题会困扰咨询师。”

青桦说：“那您不能克服一下吗？”

我说：“不能啊，青桦。心理咨询师是人不是神，凡人所具有的弱点他都可能具有。咨询师必须遵守的所有规章守则，都是付出了沉重的代价甚至包括鲜血才换来的，这就像电工在维修电器之前，必须要保证电源是关闭状态。虽然简单，但违背可能致命。”

青桦说：“好吧，我知道了不给熟人作咨询，是咨询师的守则。这对咨询师不好，但您刚才说这对来访者也不好，我却有点想不通。比如我找您咨询，我跟您熟，信任您，心里一点也不害怕不畏缩。这对于建立关系不是非常好的开端吗？”

我说：“青桦，你还挺懂行的。咨询是一个整体的过程，并不仅仅限于开头。照你说的，因为熟，所以你不怕我。这诚然不错，但进行下去，我说的话你也可以因为熟悉，就置若罔闻。我没有法子维持咨询中应有的张力，也不能确保自己中立的态度。无法全心全意地来帮助你成长，这不就是对来访者最大的不利吗？”

青桦说：“嗨！想不到因为我妈妈和您是朋友，我就没有机会找您咨询了。”

我说：“天下的好咨询师很多。我这就给你介绍几位。”说着，我就开始写给她咨询师的地址和电话。

The daughter of an old friend, Chinghua, called saying that she wanted to consult me over her leftover state. I refused, as I do not offer consultation outside office hours, and never counsel acquaintances. It was against the rules of my profession and bad for both sides. Chinghua and I finally agreed that I would treat her as the child of an old friend— actually I had known her since the day she was born.

When we met, Chinghua suggested that we revert to the old days when I would advise and she, being child of my old friend, would listen—or not.

That was fine by me and we sat down to business.

青桦说："那如果我不是咨询，只是想从你是我妈妈好朋友的角度，听听您的意见，你可以说吗？"

我说："那当然可以。你不要把这当成是一次咨询，我也不把你当成来访者。咱们聊聊天。青桦啊，那你可就要听好了。"

青桦说："哎呀，看您这样子，可比咨询时厉害多了吧？"

我说："今天我的身份，就是你一阿姨，是从小看着你长大的长辈。我说的话，有可能不客气。你如果不乐意了，可以起身就走。"

青桦说："洗耳恭听。"

我说："如果你不打算成家了，这是一个重大的决定。因为你不但要对抗世俗的眼光，你还要对抗自己的荷尔蒙。走一条和别人不同的路，和大多数人不同的路，这是一个严峻的挑战。我希望这是一个人清醒成熟的主动选择，而不是一拖再拖，成为被动局面的一个遁口。两者的结果看起来似乎是一样的，都是一个女子孤身走过一生，但实际上的感受会大不相同。"

青桦说："我只是一个寻常女子，我还是希望按部就班地走完普通女人的一生。"

我说："第二个是请你放下幻想，准备斗争。"

青桦吓了一跳，说："我和谁斗争啊？"

我说："别这么紧张，和自己斗争。就是说，你对婚姻，不要抱有太多的梦想。它是一种亲密关系，没有血缘濡养的亲密关系。它比你以前经历过的所有事情都复杂，你要做好充分的心理准备。要有耐心和恒心，要有勇气和责任。"

青桦笑起来说："这可真不像心理医生了，像我妈。我妈就这样说个不停。"

I came to the point directly by saying:

"First, if you choose to remain single, you are making a serious decision—to be different, to avoid the beaten path, to overcome the urge of your hormones. This must be a considered choice, not a lame excuse when things have dragged to a point beyond your control."

I pointed out that it may seem to be the same thing to the world—you are single, and that's the end of it. "But to you yourself," I insisted, "it makes a great difference—whether you have remained single through a conscious decision, or that your lonely life had been dragged out through force of circumstances."

Chinghua said that she was an ordinary woman and wished to have a life like any other woman.

"If that is what you want," I answered, "then the second item on the agenda is to give up all illusions and fight for what you want."

Chinghua was shocked. "Fight?" she asked, "Against whom?"

I assured her: "Don't be shocked. I mean fighting against yourself, which is to say, getting rid of illusions about marriage. Marriage is a close relationship unconnected by blood, more complicated than anything you had gone through in your life so far. You must be well prepared. You must have patience and determination, as well as courage and a sense of responsibility."

Chinghua laughed out loud: "You don't sound like a counselor. You sound like my mom, going on and on with endless advice."

我说："我不会说个不停，只说这一次。但这个世界上有什么紧要的好事情，不需要耐心恒心勇气和责任感呢？有些话，因为说得太多，就失去了新鲜感。但爱情和家庭都是很古老的东西了，你不要太期望新鲜，还是相信古老吧。"

"第三点，"我说，"那就是你要抓紧。记得有位伟人说过，抓而不紧，等于不抓。在婚姻这件事上，既然不打算独身到老，就要积极行动起来。"

青桦说："那我岂不变成了花痴？"

我说："别打岔，听我说。我可不喜欢小孩子插科打诨故作幽默。"

青桦说："你的来访者对咨询者也会这样说话吗？"

我说："一般不会。看心理医生是一件严肃的事情，来访者知道分散精力，浪费宝贵时间，得不偿失。时间都是他自己的，是用金钱买来的。当然如果他一定要这样表达，我会和他讨论。真正的花痴是什么意思，请他说明白。很正式的指示他，通常并不调侃。"

青桦撇撇嘴说："哎呀，对花痴，我也不很了解。只觉得武侠小说中一碰到傻呵呵爱恋的女子，见了帅哥围追堵截不放手，就会被人这样骂道。"

我说："你知道我是当过几十年医生的人，依我的医学知识，'花痴'其实是中医的一种病名，也就是现代医学所说的'性欲亢进'。无论男女，如果对性行为要求过于强烈，成为一种疾病的话，大夫们就会下这个诊断。主要症状有：性兴奋出现频繁，性要求异常迫切，性生活频率增加，房事时间延长……"

青桦吓得直眨巴眼睛，说："呀，好吓人！再不敢乱用这个词了。我不是花痴。"

"Listen," I retorted, "I am just saying this once. Is there anything desirable in this world that could be achieved without patience and courage and determination? Repetition is boring, but love and marriage are as old as the hills. Don't trust the new-fangled stuff. Stick to the old."

"Item three," I continued, "Since you do not choose spinsterhood, start looking for a spouse!"

"Won't I be seen as a sex maniac?"

"Listen, don't interrupt. Are you trying to be funny?"

"I don't know what I am talking about. But in the old kunfu novels, girls who go after the heroes are dismissed as sex maniacs."

"I had practiced as a doctor for many years. As far as I know, in Chinese medicine the term refers to a specific disease. In modern medicine it refers to people who are over-sexualized, and are more demanding than normal."

Chinghua rolled up her eyes in mock horror. "Gosh, I will never use this term ever again! And I am not a sex maniac!"

我说："那不要给自己乱扣帽子，武侠小说并不是真实的生活。咱回到'抓紧'上面。"

青桦说："你说我怎么抓紧呢？"

我说："这可就是你自己的事了。不要推诿责任，把这事推给父母，推给缘分，推给老天爷，推给我这样的人……这不是我们这帮人的事儿，是你自己的事儿。"

青桦说："还有第四条吗？"

我说："没了。已经够多的了，如果你能把这三件事做好了，这事儿基本上就解决了。"

青桦说："毕阿姨您再想想，还有什么可补充的？"

我说："没有补充的，但有提醒的。"

青桦说："关于对方的家世？人品？身高？职业？学历？长相？籍贯？"

我说："都不是。这是你考虑的问题，我哪里知道？我要提醒你的是价值观。"

青桦大笑道："您现在倒是真的不像心理师了，像我们学校的政治辅导员。"

我说："青桦，看来你真是没有做过正规的心理治疗。其实，心理师和来访者在咨询室里，最常讨论的就是价值观问题，只是可能用的不是我这种语气。"

青桦说："阿姨，谈对象，怎么谈到价值观上了？好像在上一堂马哲政治课。"

我说："并不是只有马克思主义哲学才讲价值观，封建主义资产阶级照样有价值观，而且无所不在地渗透到各个领域。两个相爱的人，如果在生物属性上特别相宜，但在价值观道德观上剧烈冲突或者干脆背道而驰，那么恋情可以一触即发，但婚姻常常饱含危机。这就是我要给你的提醒。"

"Don't use fancy terms indiscriminately. Let's get back to the business of looking for someone suitable. Seize the moment..."

"How?" Chinghua interrupted.

"That is up to you. Don't pin it on your parents, or trust to luck, or God's will, and don't pass it on to people like me. It is your own business."

Chinghua asked, "Is there a fourth rule?"

I said, "No, no more. Three rules to go by are quite enough."

"Please think hard? Have you anything to add?"

"Well, there is something I must remind you of...."

"You mean inquiries regarding family background, character, height, job, education, looks etc.?"

"None of the above; those are your own concerns. What I have in mind are values. What is your system of values?"

Chinghua laughed: "Now you don't sound like a counselor. You sound like the political instructor at school!"

"Values are not limited to political instructors. All social systems uphold their own system of values which permeate every aspect of life. A pair of lovers could hold vastly different moral values. In which case, they may be physically compatible but their marriage would be rocky. This is my last reminder to you."

青桦是个聪明的姑娘，频频点头。我不知道这些话她真明白了，还是出于一种礼节。

过了一段时间，青桦开始恋爱了。又过了一段时间，青桦说她要结婚了。她来给我送喜帖，说：“毕阿姨，我愿意你把我的故事写出来。”

我说：“你有什么故事啊？很多人都觉得自己的故事值得写出来，其实，有点自恋。多半高估了自己，你经历的基本上算是平常女子的简单问题。”

青桦说：“给剩女看啊。人们以为剩女出嫁很难，其实并没有那么难，我能现身说法。”

我说：“世上无难事，只要肯登攀。这事还不用登攀，只要眼睛向下一点，目光放久远一些，就走出自己的路了。”

Chinghua nodded vigorously, though I am not sure whether she understood, or was just being polite.

After a while, Chinghua started dating.

After another while, the wedding was on.

She delivered the invitation card to me personally, saying "Auntie, I would like you to write down my story."

I said, "Your story? What story? Some people think their own lives are unique and worth writing down. They are just being self-centered. What they went through is something that all women go through!"

"Write for the 'left-over' women. My story will tell them that it's not so hard to find someone."

"There is nothing on earth that cannot be done by trying. And in this case, it is not hard at all. All you need is to keep your eyes to the ground, but train your sights into the distance!"

心理测验的批发商

常常听到朋友们说，嗨！我刚做过了一个心理小测验，分析结果说我是怎样的人，实际上我并不是那样的人。从此，我就不相信心理学了。觉得尽骗人，和江湖上算命的差不多。

也有朋友说，我做过一个小测验，那真是太准了。以后，我只要看到报刊杂志上有这类文章，都会兴致勃勃地拿来一做，还迫不及待地推荐给别人。好玩不说，真是灵验啊。

这类大致以“看穿你的心”为名的小测验，如烂漫山花，弥漫四周，类似巫师发出的咒语，具有蛊惑人心的魔性。

我从来不做，不是因为斟酌它灵或是不灵，只是觉得一门严肃的科学，被随意拿来消遣，如同殷墟的甲骨，砸碎了煎汤，太轻慢。

有的测验，说你想象自己正在画画。画的是什么？国画？油画？山水风景？美人佳肴？萝卜白菜？信笔涂鸦？抽象挥洒？你可知它们说明了什么？

A Wholesale Business in Psychological Testing

Friends often talk about their experience of 'instant psychological testing." They say that the results would be totally beside the point, miles away from the kind of person that he (or she) really was, that the 'test' they took was just a hoax of wandering fortune-tellers, and that they themselves have totally given up on so-called psychological testing.

Other friends, however, would tell me: "I took a little test, and guess what, it really worked. The answers that came up were correct to the last detail!" So they would recommend those psychology games to their friends. "It not only worked," they would say, "but it was fun."

Nowadays these little tests, often under such titles as "Seeing Right Through You" are a common feature of the daily papers. Like a witch's magic, this kind of testing seems to cast a spell over its fans.

I never take part in such games, not only because I am dubious of their effectiveness. I feel that making a game out of a branch of science is frivolous, even sacrilegious, like breaking up ancient oracle bones to boil soup.

Some psychological tests kick off on the assumption that you imagine yourself to be painting. What are you painting? Traditional Chinese ink painting? Western oil painting? Painting of a beautiful lady seated in front of a banquet? Painting of cabbages and turnips? A spontaneous spray of color? An abstract scrawl? Do you really have a clue as to what these "paintings" are trying to express? And so on and so forth.

有的测验，假设大家正在等电梯。你是一直仰头看着表示电梯层数的数字？还是不耐烦地频频揿着按钮？要不干脆利用这个时间，欣赏一下同样苦等电梯的美眉的超短裙？

人们充满了好奇。就算有人对外部世界不好奇，对自己也难逃好奇之箭。谁不想知道在三千烦恼丝包裹之中的颅骨下，栖息着怎样的奥秘？它在暗中支配着你的一颦一笑，操控着你的命运舵轮，你不能对它一无所知。假若年老，生命之纸已然破旧，涂了很多若明若暗的图谱，余下的天头地角也不宽裕了，不找也罢。年轻人则更希望多了解自己，未来对于他们，具有更柔软的可塑性。

某天，碰到一位美丽女子，长发飘飘。她妩媚一笑说："我和您是同行。"

我这半辈子从事过好几种职业，一时不知道她指的是哪一行，问："你是军人吗？是内科医生吗？或是写作？要不你开了心理诊所？"

她笑笑说："都不是。"

我纳闷道："那咱们同的是哪一行呢？"

她说："我编心理小测验。"

我说："原来报刊上登的那种心理小测验，都是你编出来的。"

她很谦虚地说："不敢当。哪能都是我编出来的呢？我一个人没有那么大的能量。"

我说："你在哪里读的心理学课程呢？"

她第三次笑了，说："我没有读过心理学课程。如果我真读了相关的课程，很可能就不敢接这活儿了。"

我纳闷："你的这种测验，是怎么编出来的呢？"

Other tests make you visualize yourself squeezed in a crowd waiting in front of a lift. Are you straining your eyes to follow the numerals which denote the stops of the moving lift? Are you pressing the button impatiently? Or perhaps you are peeping at the legs of the young lady squeezed in the crowd?

People are curious. Even if they are not interested in anything outside their own circle, they cannot help being curious about themselves. They want to know the secret spring of their "mortal coil" —that which controls their glance, their smile, their destiny. So they get hooked by these tests.

When overcome by age, the sheet of your lifespan is scratched and tattered; there is not much space left, and not much worth looking for. It is the young who want to know more about themselves—their future is malleable.

One day I met a beautiful young woman with flowing black hair. Smiling sweetly, she said, "We are in the same line of business." I have been involved in many different "lines of business" and did not know which she was alluding to.

I asked: "Oh, so you have been in the army? Or a doctor of internal medicine? Or a writer? Or a counseling psychologist?"

She said, smiling: "None of the above."

I was intrigued. Now what can we have in common, I wondered.

"I am a compiler of psychological tests," she confessed.

"Ah! So the little psychology tests in the back pages of the dailies are your handiwork?"

She said modestly: "I can't do everything myself."

I asked: "Where did you take courses in psychology?"

Smiling, for the third time since we met, she said: "I have never taken courses in psychology. If I did, I probably would not dare to take on this job."

I was intrigued. "Your psychology tests, how did you manage to concoct them?"

她看了看四周，很神秘地说："如果是别人问我，我就不告诉他。因为尊敬您，所以，全盘告知。"

我一下子有点紧张。凡是听到人谈到秘密的时候，第一个反应就是想上厕所并且有点害怕。要是将来一旦秘密泄露了，岂不要担干系？

美丽的女子款言道："您不用怕，其实这也是半公开的诀窍。一般的人，以为是先编好了测验的故事，再来确定答案，其实，不然，是先设计好了不同的人会有怎样不同的反应，然后再来设计前缘。"

我说："能举个例子吗？我还是不大明白。"

美丽女子说："比如，人们面对突然的巨响，会有不同的判断和应对模式。谨慎而且惜命的人，首先想到的是安全问题和自保；勇敢和喜爱助人的人，首先想到的是一探究竟和挺身而出；教条和僵化的人，很可能麻木和迟钝，不能审时度势；胆小如鼠的人，当然是惊慌失措和打哆嗦了。

"你先把各种人不同的反应方式找到，然后再反推回来，设计出相应的情境，不是就水到渠成了吗？你顺势即可编一个心理小测验：

春天，你和朋友们正在郊外空旷的草地上用餐，突然电闪雷鸣并且听到野兽的吼叫，你会采用哪种方式：

She looked around and put on an air of mystery, whispered: "I never tell this to anybody. But you, you are different. I respect you. You will get the whole story."

I was alarmed. Whenever people start telling me their secrets, I get scared. I tense up and head for the bathroom. If the secret is out one of those days, will I be dragged into the mire?

The young woman said slowly, "You need not worry. The whole operation is actually an open secret. Most people assume that the questions in the psychological tests are formulated first, expecting to be followed by the correct answers. Actually, it is the opposite. We anticipate the possible answers from different kinds of people, and we design the questions correspondingly!"

I was still confused and asked her to give me an example.

The pretty young woman obliged: "Supposing suddenly there was a loud bang. People would have different interpretations of the sound and react accordingly. The timid type who cares only for their own skin would worry about their own safety. The manly type would try to find out what has happened and think what they could do to help. Those who are completely dumbstruck cannot react in any sensible way. The cowards, of course, would be trembling with fear."

"After laying out all possible reactions," the pretty young woman concluded, "We would go backwards and design the corresponding situations which would elicit those expected reactions."

She added, "Actually you can design a psychological test at a snap of your fingers. Say it is spring. You and your friend are enjoying a picnic in the suburbs. Suddenly there is thunder and lightning and the howl of wild beasts. What would you do?"

She answered herself:

A．堵起耳朵，哭泣，瘫倒在地。

B．用身体掩护朋友，说，不要慌，有我呢！说着拿起一根粗壮木棒，警惕地四处巡查。

C．一句话也不说，撒腿就跑，看到不远处有一个土坑可以藏身。

D．抬头看看天，佯作镇定地说，临来之前我查了资料，天气晴朗，这一带没有大型野兽，不必害怕。

“按照刚才咱们前面说到的逆推理法，相应的分析，很容易完成，不过举手之劳。”

我目瞪口呆，说：“就这么容易？”

美丽女子说：“这还算比较复杂的呢。有时候，简单的心理小测验，我一天能编出十多条呢！一条能赚几百块钱，你可以算算收入。我真要感谢喜欢心理学的人，他们爱看，报刊才会登，我拿了稿费，才有余力买漂亮的裙子。”

我试探地问：“如果我把你的创作过程告诉更多的人，你会不会断了生意？”

她爽快地说：“不会。总有人喜欢神秘又无法验证的东西，我就是一个心理测验的批发商。”

"A, stop my ears, cry, crumble up in a heap on the ground;

"B, shield my friend with my own body, saying 'Not to worry, I am here,' and then take up a stout club and walk around, safeguarding my friend;

"C, seeing a hiding place in the distance, dash for it without saying a word;

"D, look up at the sky, put on an air of confidence and announce 'I just checked the weather, it is fine. There are no large animals in this area. No need to worry.'"

The girl concluded: "By moving backwards as I had just explained, it is easy to deduce the question and concoct the right answer. You can do it with your eyes closed..."

I was dumbstruck, "Just like that?"

The girl scoffed: "What you just heard is a sample for a complex case. As for the easy kind, it is just a little question and answer. I turn them out by the dozens every day and get paid one hundred yuan for each item. You can figure out for yourself how much I make per day. I am so grateful to the people who like to dabble in psychology tests. They love my questionnaires. The dailies print them, and I pick up the money. The money pays for my pretty skirts."

"If your operation is exposed, your cash flow will dry up," I warned.

She smiled and stated: "No fear of that. People are charmed by that which is mysterious and cannot be tested factually. I am a wholesale dealer in psychological testing."

心理测验的批发商

A Wholesale Business in Psychological Testing

要以自己的真实面目示人，没有必要取悦他人，没有必要委屈自己。

适度撒谎

千万不要说谎。所有的大人们都这样说。

但这实际上就是一句谎话。几乎所有的人在他的一生的某些时候都曾说过谎，谎言就像毒药，有时候非用不可。

如果笼统地说不能撒谎，有人就会在暗地里偷偷地笑。因为这实际上是做不到的。就连列宁那样的革命导师，小时候打碎了花瓶都不承认，可见这真是人类的普遍弱点。

那么怎么将谎言的危害减到最小呢？这就好比是抽烟，抽烟自然是不对的，但全地球的人都在抽，生活在现在的人们，就在大力提倡无烟环境的同时，也生产加长过滤嘴和减毒的香烟，使吸烟的弊病小一点。

既然完全不说谎几乎是做不到的，我们就应该把说谎局限在一个狭小的范围内，就好像世上有个谎言国，如果我们万不得已要在其中穿行，千万不要越过危险的界碑。

怎么掌握这个界限呢？

我觉得必要的撒谎首先是要为了一个善良的目的。比如我当过许多年的医生，有的病人得了不治之症，但他们还是非常想活下去。他们常常拉着我的手问："您说我的病还治得好吗？大夫！"我会毫不迟疑地告诉他："能治好！一定能治好！"这就是善良的谎言。

Discreet Lying

"Whatever you do, never lie!" This is what we are told by adults, though impossible to carry out in real life.

Everybody, at some point in his or her life, must have lied.

Lying is like taking poison—sometimes you can't do without it.

Now if you state categorically that lying is forbidden, period, somebody will be secretly laughing because it is asking for the impossible. Even Lenin, the revolutionary leader, was not exempt. As a child, he had broken a vase and lied about it. Lying, it seems to be a weakness that no one is free from.

If this is so, then how can we minimize the harm? Take smoking, for instance. The whole world knows that smoking is bad, but the whole world keeps smoking. People mark out non-smoking areas; they manufacture long-stemmed nicotine-free cigarettes and what not, all for the purpose of reducing the harm of smoking. But they keep smoking. So it is with lying.

Since it is impossible to completely wipe out lying, we should limit it to a minimum. It is as if there is a kingdom of lying: if you have no choice but to cut across the grounds, be sure not to stumble into the danger zone.

The point is, how do you keep within limits?

Well, for starters, if a lie is unavoidable, it must be for a good reason. I had been a practicing physician for many years. Some of my patients, victims of incurable disease, would hold my hand and ask if they will be alright, and I would always assure them that they would. Well, I would call that a well-meant lie.

但有一个小朋友考试成绩不好，他就对父母撒谎，胡说一个好成绩或者干脆不告诉家长。你问他为什么要这样做，他说是怕爸爸妈妈生气。这应该说是一种善良的瞎话，可不知他们往深处想了没有。学习成绩不好，自己没本事赶上去，藏着掖着，就会越落越远。这不是等于把匕首扎得更深，在更大程度上伤了父母的心吗？你说你这话到底是善意还是恶意呢？

善意和恶意的谎言的界限并不是那么容易分得清的。有的事，你瞒得了一时，瞒不了一世。从长远看，一般的情况是晚说不如早说，实事求是最简单。所以，你要是把握不住善意和恶意的界限，还是说实话的好。

第二个界限是适度的谎言必须没有恶劣的后果。换句通俗的话说，就是这个谎话无伤大雅。比如，一个啰唆的朋友打电话来，重要事情说完，还没完没了。我又不好意思打断她，就说：“哎呀，对不起，我马上要出去办个事，今天就谈到这儿好吗？”对方就很体贴地把话筒放下了。这是不是谎言？当然是的。但我想它没有什么不好的结果，既节省了时间，大家又都有个台阶下，所以这个谎是可以说的。这种推辞的话，有的时候被称为委婉。委婉其实就是一种适度的谎言。

但同学们到底见的事比较少，经验还不足，有时候估计不出自己的谎话有多大的危害。因为这需要有一定的社会经历，有一定的预见性。如果大家还不具备掌握分寸的能力，我看为了保险起见，还是说实话的好。

But supposing a little boy did not do well at his exam and lied to his parents about his grade, or decided to keep mum about it. Now why did he do that? The boy claimed that he did not want his parents to be upset. Thus his lie might be justified as a well-meant lie. One wonders, however, if the boy had thought ahead—that by lagging behind and putting his problems behind him, he was digging himself deeper into the mire, and eventually inflicting greater pain on his parents. Seen from this angle, could the boy still claim to be telling a well-meant lie?

There is no clear line of demarcation between a well-meant lie and a lie of evil intent. With all the best intentions in the world, it is hard to keep up a pretense for long. Generally speaking, it is advisable to "pour it out" early rather than late. Honesty is the best policy, as the saying goes. To sum up, when unsure of the pros and cons of the situation, it is advisable to stick to the facts.

The next principle to hold on to is that an unavoidable untruth must not carry with it any harmful side-effects. Spelled out in plain speech, it means that an untruth must not be guilty of any impropriety. Supposing a friend phoned and after delivering her message, started off on a marathon chat. Not wishing to cut her short abruptly, I would apologize and say that I have business to attend to, and that we would speak later. She at the other end would put down the receiver. Have I been lying? I certainly did, but I did no harm. Nobody lost face and I saved time. So mine was an acceptable fib. This kind of fibbing is considered tactful. Actually, tact is by nature a partial evasion of the truth.

Generally speaking, students are young and inexperienced and might not realize the extent of the harm caused by their lying. For that, one needs more experience of life as well as foresight. Since one cannot expect school kids to observe the niceties of socializing, I recommend that they play safe and stick to the facts. "Honesty is the best policy."

适度谎言的最后一条原则——它必须是为了维持人的自尊心而说。考查普通人们说谎的动机，许多时候都是为了保存自己的脸面。这是可以理解的。俗话说，人有脸树有皮嘛！如其在众人面前被强烈地践踏了自尊，搞得长久抬不起头来，不如暂且不吐露事情的真相，维持形象的完整。当然这绝不是原谅自己的缺点错误，而是要在人群背后，痛下决心，深刻地检讨自己的所作所为，从此再不犯同样的过失。这说起来容易，其实挺难的。因为没有人监督，很容易又走上老路。假如你没有足够的自制力，没法保证自己能暗暗地改正错误，还是公开承认的好，这样虽说暂时丢了面子，从长远看，还是有益的。最怕的是自己没有勇气承认，又没毅力改正，就会越滑越远。

我们今天谈论了一个非常敏感的话题。我的本意绝不是倡导大家来说谎，或是说谎言无害。不，谎言是非常有害的。它会使我们的良心背上沉重的负担，从此失去天真快乐。它会带来意想不到的灾难，把我们推进黑暗的深渊。少年朋友，你要记住，在所有能够说真话的时候，请一定说真话。在万不得已说假话的时候，你必须担保它没有恶劣的后果。假如它涉及你的过失，你在维护了自己尊严的同时，一定永不再犯。

人的一生像一次艰苦的航行，谎言国几乎是必经的黑海洋。操起你的双桨，勇敢而又策略地穿过去，千万不要在风浪里翻了你的小船。做一个诚实谦虚、知错就改的人，你就有希望到达金色的彼岸。

The last principle of discreet lying—it should be allowed as the last line of defense of personal dignity. Generally speaking, people lie mostly to save face. This is understandable. As the saying goes: "As the bark to the tree, so the skin to a man's face." Thus it is understandable that a man would hide the truth when he is involved in scandal and his personal dignity and reputation is dragged in the mire and he cannot expect a quick recovery. To avoid such a fate and save face, he would rather suppress the facts and lie low for the time being. This does not mean that he is clinging to his offence. Ashamed to face the world, he is repentant for his offence and determined to make a new start. This sounds easy, but without oversight, he is liable to slide back into repeating the offence. If one does not have enough self-control, and is not confident of changing one's ways in secret, then it is advisable to come clean in public. It entails loss of face for the moment, but in the long run, it has its advantages. The worst scenario is when one has neither the courage to acknowledge one's fault, nor the will power to change for the better on one's own initiative—such a person will go down in iniquity.

We have been talking about a very sensitive issue. I am certainly not proposing that everybody start lying, or to claim that lying is harmless. On the contrary, lying is very harmful. It is a burden weighing down our conscience, causing us to lose our innocence, bringing in its wake unexpected disaster, dragging us down in iniquity. Young friends, please hear me out and remember: always tell the truth! Only under extreme circumstances and when there will be no harmful consequences can you afford to tell an untruth. If the case affects you personally, you must make sure that there will never be a repeat offense.

A person's life is like an arduous trip, and the black sea of lies is spread across your path. Pick up your oars and cross this sea of lies with courage and dexterity. Keep your little boat steady in the roaring waters. Be honest, be humble, dare to correct yourself when you are wrong—and you will be able to reach the golden shore.

分泌幸福的“内啡肽”

我曾看过一则新闻：英国有家报社，向社会有奖征答“谁是最幸福的人”，然后排出第一种最幸福的人，是一个妈妈给孩子洗完澡、怀抱着婴儿；第二种最幸福的人，是一个医生治好了病人并目送他远去；第三种最幸福的人，是一个孩子在海滩上筑起了沙堡；备选答案是，一个作家写完了著作的最后一个字，放下笔的那一瞬间。

看完这则不很引人注目的报道，那一瞬间，我真的像被子弹打中一样，感到极度震惊——这四种状况都曾集于我一身，但是，我没有感觉到幸福！

我为什么没有幸福感？有了这个问号后，我就去观察周围的人，这才发现，有幸福感的人是如此之少。有一年，我拿出贺卡看了看，结果发现最多的是“祝你幸福”，这可能是中国人的集体无意识，所以才会觉得是永远的吉祥话。

可是，幸福的本质是什么东西呢？

Secretions of Happiness

I once came across a word game, set unobtrusively in a corner of an English newspaper, offering a prize for the best answer to the question "Who is the happiest person on earth," listing four answers to choose from:

1. a mother holding her baby after giving it a bath;

2. a doctor seeing off a patient who had recovered;

3. a child who had succeeded in building a sand castle on the beach;

4. a writer putting down his pen after writing down the last word to his book.

The fourth was the correct answer.

As I put down the paper, I felt as if I had been hit by a shot—I have gone through all the experiences described above, and have never felt any happiness at the time!

Why had I not felt any happiness at those moments? With this question in mind, I began to observe the people around me and found that very few people are conscious of being happy. As I ran through my collection of greeting cards at the end of the year, I found out that most of them wished me happiness. Those auspicious words must be the collective unconsciousness of the Chinese people, to be forever wishing for happiness.

But what is the nature of happiness?

日本春山茂雄博士的《脑内革命》一书说，当我们感知幸福的时候，其实是生理在分泌一种内啡肽，即幸福感是体内内啡肽的分泌。从罂粟里提炼的吗啡是毒品，它的魔力正是在于它的分子结构模拟了生理基础上的内啡肽，让你体验到一种伪装的、模拟的快乐。当你觉得真正快乐的时候，例如接到大学录取通知书时，如果去抽血查验体内的生化水平，你的内啡肽水平是增高的。

据春山茂雄研究，人体内啡肽的分泌，和马斯洛“需要层次”的金字塔理论惊人吻合：吃饭能带来愉悦，人在生理基础上是快乐的；然后，在实现安全、爱和尊严的需要的过程中，伴随着更大量内啡肽的分泌，让你感知自己的幸福；最重要的是，当你完成自我实现的时候，内啡肽就达到非常高的水平，远远超出吃饭带来的幸福感。

这种生理和心理的结合，使我觉得，能够体验到幸福感，是一个需要训练、感知且不断提高的过程，因为为幸福不是与生俱来的。

我觉得世界上的幸福，首先来自一个坚定的信念。

我常去高校和大学生交流，给我最多的感觉是，他们面临一个非常重要的问题——人生观的确立和价值观的走向，即人为什么而活着。

The Japanese scholar Shige Haruyama in his book *Revolution Inside the Mind* had said that when we feel happy, it is actually a reaction from secretions of our endoctrinal glands—the source of our feelings of happiness.

Secretions from the poppy seed is a drug. Its power lies in the fact that its molecular structure is an imitation of the physiological basis of endorphin, exuding a sense of happiness. Supposing you are truly happy (receiving a letter of acceptance from the college of your choice), you will find that your level of endorphin is higher than normal if you get a bio-chemical levels check-up.

According to the research of Shige Haruyama, human secretions of endorphin confirms, to a surprising degree, to Maslow's "pyramid" theory regarding the human "levels of necessity." For instance, having dinner makes one happy—on a biological level. In the course of achieving your needs for safety, love and respect, you will enjoy cumulatively higher levels of secretions of endorphin. Finally when you have achieved self-realization, you will reach an even higher level of endorphin, bringing greater sense of happiness, far beyond the happiness of a good dinner.

This theory of the combination of the physical and the psychological made me realize that achievement of happiness is a continuous process of training, feeling and heightening of consciousness, because human happiness is not innate from birth.

Personally, I believe that human happiness comes first and foremost from personal convictions.

I often meet with college students, and I feel that they are all faced with an important question—their outlook on life and the direction of their scale of values, which is in fact, the question "What are we living for?"

经常有媒体采访我的心理咨询中心，最喜欢提的问题是：“咨询最多的问题是什么？”我说：“心理咨询室这张米黄色的沙发如若有知，一定会一次次地听到来访者在问‘我为什么而活着？’”

我觉得人是追索意义的动物，尤其是年轻人，都曾经无数次地叩问过这个问题。

以前，我们喜欢用灌输式的方法，从小将主义、理想或目标灌输给孩子，希望能够在他们心中扎下根，成为他们一生的坐标。可我现在发现，一个人的目标，一定需要他自己经过艰苦的摸索，然后在心理结构里确立下来，否则，无论我们多么用心良苦、谆谆教导，它真的只是一个外部的东西。

其实，每个人都早早地确立了一生的目标，因为它原本已存在于你的内心：从童年经验开始，你所热爱、尊敬、向往、要为之奋斗的东西，其实早已植根于心里，只不过被许多世俗的东西、繁杂的外界所影响，甚至被遮蔽了。当一个人开始有意识地关注自己的心理健康，那是在清理他的心理结构，然后明白心中取得最主要作用的架构和体系。

我曾在一所非常好的大学做讲座，台下有学生递条子说：“毕老师，我想问问你，我年轻貌美，又有这么好的大学文凭，要是不找一个大款把自己嫁了，我是不是浪费了资源？”我想，在大学生寻找目标的迷茫过程中，能够有这种朋友式的探讨，是特别重要的。

另外，我觉得自我形象的定位是幸福感来源非常重要的一部分。

Members of the press sometimes visit at my psychological counseling center and one of their favorite questions used to be: "What is the most recurrent problem that you are up against?" And I would tell them: "If this sofa here could speak, it would tell you that what it hears *ad nauseum* is the question: 'What am I living for?' I believe that the human being is an animal seeking for a meaning to life, especially the young, who are perpetually asking this question."

In the past, we resorted to the "brain-filling" method, talking to kids about isms and ideals and targets, hoping that these would strike roots in their little hearts and guide them for life. But by now I have discovered that setting up a goal for life is a process achieved through personal exploration, and ultimately instilled into the psychological system. Otherwise, no matter how hard we try lecturing and hectoring, it will always be something alien to the self.

Actually everybody has a target in his or her heart, imbued into the consciousness since early childhood. Your preferences, your yearnings, the goals that you are ready to fight for, have long been instilled into your heart, but smothered by worldly concerns and alien influences, or clouded over. When you start being concerned over your psychological health, you are actually doing an overhaul of your psychological make-up and getting to know the frame-work and system dominant in your own heart.

I was once giving a talk at a prestigious university, and one of the students passed up a slip of paper asking: "Dear Teacher Bi, I have a question for you. I am young and good-looking, and about to get a degree from this prestigious university. Now if I do not go and marry myself off to a big shot, won't I be wasting my resources?" I believe that such frank discussions are very important to college graduates who are fumbling in the dark regarding their goals in life.

I also believe that one's self-image is an important part of one's source of happiness.

在大学生自我形象的构建里，有一部分是他们的“出身”（阶层）：他们从各种阶层突然聚合到一起，大学虽是个相对小的、封闭的环境，却也是整个社会的缩影，因此，如何看待自己不可选择的出身阶层，这是自我形象非常重要的部分。另外一部分是他们的学业，包括学习的能力、智商的能力、人际交往的能力等，可归为自己奋斗来的部分。

然而，还有特别重要的一部分，就是外在条件——长相。

我曾在一所大学做关于自我形象、自我认知的讲座，请台下的学生回答：你们有谁曾经为自己的长相自卑？结果齐刷刷地举手——所有的人都自卑！

我当时一下子不知该如何反应，没料到当代年轻人在相貌问题上居然有如此大的压力。

后来，我悄悄问一位女生，问她为自己相貌的哪一点自卑，我实在找不着——她身材窈窕、黑发如瀑、明眸皓齿、肤如凝脂，真的是美女。

她说：“我有一颗牙齿长得不好看。”

我说：“哪颗牙齿？”

她说：“第六颗牙齿。”

我说：“谢谢你告诉我，否则站在对面看你一百年，我也看不出你那颗牙齿不好。”

她说：“你不知道，可是我知道。我不敢笑，从来都是抿着嘴只露出两颗牙齿。同学都说我多‘冷’、多高傲，其实，我只是怕人看到第六颗牙齿。男生追求我的时候，我就想，我一颗牙齿不好，他还追求我，肯定是别有用心，于是放弃了好几个条件很好的男生。”

One component of college students' self-image is their family background (including class background). Coming from vastly different class backgrounds, they now form a new entity. Though a small, closed environment, college is still a condensed version of the world at large. Thus the problem of how you deal with your class background (something utterly beyond your own control) is an important part of your self-image.

Other parts of your make-up include your scholarship, including your performance in class, your ability to deal with relationships, your IQ...all of which could be considered as self-acquired abilities.

There is another important element to one's self-image—looks.

Once I was talking about self-image and self-knowledge at a college gathering, and I asked the student audience to answer the question—does anyone here have a sense of inferiority due to his/her looks. All hands shot up. I was flabbergasted. I had no idea that this current young generation was under so much pressure regarding their looks. Later, I asked a girl what part of her looks she was ashamed of. She said that one of her teeth was crooked. I asked which teeth, and she said the sixth. I thanked her for telling me, as I would have stood in front of her for a hundred years and never seen anything wrong with any of her teeth.

She said: "You don't know about it, but I do. I don't laugh out loud. I only smile with my lips pursed, or show at most two teeth. My classmates say that I am cold, and that I am arrogant. Actually I just don't dare to show my teeth. When one of the boys chased me, I would say to myself—I have a bad tooth and he still chases me—obviously he has sinister designs. I have already turned down several boys who are the cream of the crop."

我觉得，当一个人不能接纳自己，不能和自己友好地相处的时候，他就不能和别人友好地相处。因为，他对自己都那么百般挑剔、那样苛刻，又怎能和别人有真诚的、良好的沟通与关系？

其实，我挺欣赏基督教里的说法：接受你不可改变的那一部分。我们可以列一列，像出身的阶层、长相及缺陷，这些是我们不可改变的，而我们能够去修炼、弥补和提高的，就是我们可改变的那一部分。

面对一个我们不可改变的东西，该如何对待它，每个人的答案是不一样的，而这个不一样的答案却可能深刻地影响我们的一生。比如，一个人认为他丑，就认定自己完全不会幸福了，觉得他既然这么丑，有什么权利得到幸福？一个人说他很贫寒，为什么别人可以含着银汤匙出生，而他却含着草根出生？

面对种种不平等，我常跟年轻人说，不平等是社会有机组成的一部分，而让它变得更为平等，是你义不容辞的责任之一。

首先，你要丢掉幻想，坦然接纳不公平、巨大的差异或先天不良。然后，对于自己可改变的部分，你就要细细地分析，找出自己的优缺点，是优点就让它更好，是缺点就要去弥补，尤其要突出优点，把自己光彩照人的方面表现出来。因为中国文化特别容易告诉你哪里不行，生怕你忘了自己的缺点，而你有什么优点，告诉你的人可不太多，所以要坦然接受自己的优点，将它发扬光大。

It occurred to me that if you cannot accept the person that you are, and cannot live with yourself, it follows that you cannot live with other people. The reason is obvious: if you are so picky even with yourself, how can you tolerate other people? I actually appreciate the Christian teaching to tolerate the part about oneself that cannot be changed. We could make a list: family background, looks and deficiencies since birth. The only parts that are changeable are those that have been acquired through one's own efforts.

When faced with something unchangeable, people react differently, and the various reactions may leave a lasting mark on their lives. For instance, people who think that they are ugly feel doomed, convinced that happiness is beyond their reach. Another set of people believe that they are doomed to poverty. They cannot understand why others are born with a silver spoon in their mouths, while they are born chewing grass.

Faced with this kind of inequality, I would say to the young: "Inequality is an innate part of the society. It is up to you people to fix it."

First and foremost, put aside all illusions, accept injustice, gaping inequality or born deficiencies.

The next step is to face up to your own situation: enhance your strengths and overcome your weaknesses, highlight your strengths and make them visible. It is part of Chinese culture to downplay one's own strengths and be always reminded of one's own weakness. Actually, one should matter-of-factly accept one's own strengths and advantages and put them on full display.

心理咨询中心来过一位留英硕士，月薪12万元，可他将自己说得一无是处，弄得我都心酸。我才知道，一个人接不接纳自己，其实不在于外在的条件，也不在于世俗的评判标准，而完全在于他内心框架的衡量。

我通常咨询完了不会给谁留作业，但那天我说："我给你留个作业——下星期来见我之前，你要写出自己的15条优点。"

他快晕过去了，说："我怎么能找到15条优点呢？至多也就找出一两条。这个世界上，可能只有您相信我还有优点，我父母就不相信我有优点，所有人都不相信我有优点！"

我说："你老板起码相信你有优点吧，否则怎会出月薪12万元雇你？"

他突然在这个事实面前愣了半天，然后说："噢，那我试试看。"

所以我觉得，应该去认识自己的长处，将它发扬光大，去接纳那些不可改变的东西。当你能够坦然地面对自己的时候，其实也就可以坦然地面对世界——放下包袱后，你才可以轻装前进。

费尔巴哈说过："你的第一责任是使你自己幸福。你自己幸福了，你也就能使别人幸福，因为，幸福的人愿意在自己周围只看到幸福的人。"

常常听到有人说自己不幸福，希望别人给他幸福。我想，这就是他不幸福的根源。

Once there was a visitor to the counseling center, an MA candidate returned from England, who commanded a monthly salary of 120 thousand yuan and who belittled himself to such an extent that I was on the verge of tears on his behalf. He made me realize that one's sense of self-worth is not a matter of one's qualifications or one's repute—it is entirely a matter of one's self-evaluation. I don't leave 'homework' to my clients as a rule, but in this case, I gave this gentleman some homework. "When we meet next week," I said, "please bring me a written list of fifteen examples of your personal strengths!" He almost fainted. "How can I find fifteen!" he protested, "At most I can dig out one or two. In the whole wide world, you are the only person who believes that I have something to be proud of. My father can see nothing in me to be proud of, and neither can my mother. No one can see anything worthwhile in me!" I retorted: "Your boss must have seen something in you to pay you 120 thousand a month!" He was struck by the thought and stood thinking for a while. Then he said: "Okay, I will try to carry out your assignment and make a list of fifteen examples of my personal strengths."

I would advise that we see our own strengths and try to enhance our strengths, and that we accept what cannot be changed. Only when you are ready to accept yourself can you face the world at large. You must put down your mental burden and move forward unimpeded.

Feuerbach had said: "Your first duty is to make yourself happy. Only when you are happy can you make others happy. Happy people want to see other happy people around them."

People would say: "I am unhappy, I do wish someone would make me happy!" To me this is precisely the root of their unhappiness—that they rely on others to make them happy.

倾听，是你的魅力

我读心理学博士方向课程的时候，书写作业，其中有一篇是研究“倾听”的。刚开始我想，这还不容易啊，人有两耳，只要不是先天失聪，落草就能听见动静。夜半时分，人睡着了，眼睛闭着，耳轮没有开关，一有月落乌啼，人就猛然惊醒，想不倾听都做不到。再者，我做内科医生多年，每天都要无数次地听病人倾倒满腔苦水，鼓膜都起茧子了。所以，倾听对我应不是问题。

查了资料，认真思考，才知差距多多。在“倾听”这门功课上，许多人不及格。如果谈话的人没有我们的学识高，我们就会虚与委蛇地听。如果谈话的人冗长繁琐，我们就会不客气地打断叙述。如果谈话的人言不及义，我们会明显地露出厌倦的神色。如果谈话的人缺少真知灼见，我们会讽刺挖苦，令他难堪……凡此种种，我都无数次地表演过，至今一想起来，无地自容。

世上的人，天然就掌握了倾听艺术的人，可说凤毛麟角。

不信，咱们来做一个试验。

Lend me your Ear

Once, when I was working towards a Ph.D degree in psychology, I got an assignment to write a paper on "listening."

Easy, I thought to myself. We all have a pair of ears, and unless we are deaf at birth, our ears pick up sounds the minute we are born into the world. In the middle of the night, our eyes are closed in sleep but our ears are still open and alert to all the sounds of the night. By daybreak, when the moon is down and the birds start chirping, we would be wakened, and we have to listen to the sounds of dawn whether we want to or not. Moreover, I myself have been a doctor of internal medicine. By now, my ears are bursting, filled by my patients' outpourings of pain and grief over the years. To me, listening is a cup of tea. Easy.

Only after looking up documents and doing some hard thinking did I realize that "listening" is not that easy, and that many people do not pass the test of listening properly. For instancc:

If the speaker was not as informed as we were, we would be condescending. If the speaker was garrulous and boring, we would rudely interrupt. If the speaker strayed away from the subject, we would show marked signs of boredom. If the speaker did not sparkle with wit, we would raise embarrassing questions to put him down a peg or two. I blush to remember how I myself had been up to those tricks time and time again.

In fact, hardly anyone can boast of being born with the gift of listening. Let us try an experiment.

你找一个好朋友，对他或她说，我现在同你讲我的心里话，你却不要认真听。你可以东张西望，你可以搔首弄姿，你也可以听音乐梳头发干一切你忽然想到的小事，你也可以王顾左右而言他……总之，你什么都可以做，就是不必听我说。

当你的朋友决定配合你以后，这个游戏就可以开始了。你必须拣一件撕肝裂胆的痛事来说，越动感情越好，切不可潦草敷衍。

好了，你说吧……

我猜你说不了多长时间，最多3分钟，就会鸣金收兵，无论如何你也说不下去了。面对着一个对你的疾苦、你的忧愁无动于衷的家伙，你再无兴趣敞开襟怀。不但你缄口了，而且你感到沮丧和愤怒。你觉得这个朋友愧对你的信任，太不够朋友。你决定以后和他渐疏渐远，你甚至怀疑认识这个人是不是一个错误……

你会说，不认真听别人讲话，会有这样严重的后果吗？我可以很负责地告诉你，正是如此。有很多我们丧失的机遇，有若干阴差阳错的信息，有不少失之交臂的朋友，甚至各奔东西的恋人，那绝缘的起因，都系我们不曾学会倾听。

好了，这个令人不愉快的游戏我们就做到这里。下面，我们来做一个令人愉快的活动。

还是你和你的朋友。这一次，是你的朋友向你诉说刻骨铭心的往事。请你身体前倾，请你目光和煦。你屏息关注着他的眼神，你随着他的情感冲浪而起伏。如果他高兴，你也报以会心的微笑。如果他悲哀，你便陪伴着垂下眼帘。如果他落泪了，你温柔地递上纸巾。如果他久久地沉默，你也和他缄口走过……

非常简单。当他说完了，游戏就结束了。你可以问问他，在你这样倾听他的过程中，他感到了什么。

Suppose you get a good old friend to stop by. You tell him that you are going to pour some confidential info into his ears, but ask him not to pay attention. Tell him that he could gape around, admire his own shadow, listen to music, comb his hair... In a word, tell him to do any silly thing that occurs to him on the spur of the moment, even just looking around, but warn him on no account to listen to what you are saying!

When your friend is ready to play the role, the game may start. You pick a heart-rending story and tell it in great detail. Okay, now start....

My guess is that within three minutes you will stop telling your story. You just cannot find it in yourself to go on. Faced with a fellow who had not the slightest interest in your misfortunes, you have no incentive to pour it out. You have not only stopped talking, you actually feel angry and frustrated. You feel that this friend has betrayed your trust, betrayed your friendship. You decide to distance yourself from him. You feel that it had been a mistake to be friends with him in the first place.

Now you may ask: would inattention to someone's speaking carry such serious consequences? Let me tell you in all seriousness: indeed it would! There are so many missed opportunities in life, so much mangled information, so many would-be friends who strayed apart, and even lovers who missed each other by a hair's breadth—all because people fail to listen.

Let us put aside this unpleasant game and try somcthing pleasant.

Take again, for example, the case of you and your friend. On this occasion, your friend shares with you a chapter of his heart-breaking past. Now you bend towards him. Your eyes are shining with concentration. You hold your breath, looking into his eyes. Your breath matches the rise and fall of his emotional outbursts. When he is happy, you smile back at him. When he is sad, you lower your gaze. If he becomes tearful, you offer him a Kleenex. If he remains silent, keep him company for as long as it takes. Simple.

When your friend finishes talking, the game is over. You may ask him what had his feelings been during the period when you had kept up with his moods.

我猜，你的朋友会告诉你，你给了他尊重，给了他关爱。给他的孤独以抚慰，给他的无望以曙光，让他的快乐加倍，让他的哀伤减半。你是他最好的朋友之一，他会记得和你一道度过的难忘时光。

这就是倾听的魔力。

倾听的“倾”字，我原以为就是表示身体向前斜着，用肢体语言表示关爱与注重。翻查字典，其实不然。或者说仅仅做这样的理解是不够全面的。倾听，就是“用尽力量去听”。这里的“倾”字，类乎倾巢出动，类乎倾箱倒箧，类乎倾国倾城，类乎倾盆大雨……总之殚精竭虑毫无保留。

可能有点夸张和矫枉过正，但倾听的重要性我以为必须提到相当的高度来认识，这是一个人心理是否健康的重要标志之一。人活在世上，说和听是两件要务。说，主要是表达自己的思想情感和意识，每一个说话的人都希望别人能够听到自己的声音。听，就是接收他人描述内心想法，以达到沟通和交流的目的。听和说像是鲲鹏的两只翅膀，必须协调展开，才能直上九万里。

I guess your friend will say that you had been caring and respectful, and that you had lifted him out of solitude and shone a light on his future. He will say that you had heaped happiness on him and had taken away his grief, adding that you are his best friend and that he will always remember the time spent with you.

Thus is revealed the power of listening.

Of the two Chinese ideograms for "to listen"—*qin ting*, the first character *qin* denotes a slightly forward-bending position. One would deduce that such a body position denotes appreciation and attention while listening. But if we check it up in the dictionary, we will realize that such an understanding of the term *qin ting* is missing something crucial. Actually *qin ting* means listening with concentration, literally pouring out all the energy in one's body to listen. The use of the ideogram *qin* here is somewhat similar to its use in such terms as *qin cao chu dong* (an outpouring), or *qin pen da yu* (pouring rain), or *qin cheng qin guo* (a beauty that has cities and nations at her feet), in a word, an overwhelming outpouring of energy.

This might sound a little exaggerated, but I honestly believe that one can never overrate the importance of listening. It is actually a signal of one's psychological health. For humans living on earth, listening and speaking are important signs of one's psychological well-being.

Listening and speaking are two important functions. To speak is to express one's own thoughts and feelings. People wish to be heard by others. To listen, on the other hand, means accepting other people's expressions of their inner thoughts and feelings. Thus, through listening and speaking, mutual understanding and communication are achieved. To listen and to speak up are like the two wings of a mighty crane—both wings must be unfolded for the bird to soar on high.

现代生活飞速地发展，人的一辈子，再不是蜷缩在一个小村或小镇，而是纵横驰骋漂洋过海。所接触的人，不再是几十一百，很可能成千上万。要在相对短暂的时间内，让别人听懂你的话，让你听懂别人的话，并且在两颗头脑之间产生碰撞，这就变成了心灵的艺术。

现今鼓励青年展示才智的书很多，教你怎样展现自我优点，怎样在第一时间给人一个好印象，怎样通过匪夷所思的面试，怎样追逐一见钟情的异性……都有不少绝招。有人就觉得人际交往是一个充满了技术的领域，可以靠掌握若干独门功夫就能翻云覆雨的领域。其实，享有好的人际关系，学会交流，听比说更重要。

从人的发展顺序来看，我们是先学着听。我之所以用了“学着”这个词，是指如果没有系统的学习，有的人可能终其一生，都没能学会如何“听”。他可以听到雪落的声音，可他感觉不到肃穆。他可以听到儿童的笑声，可他感受不到纯真。他可以听到旁人的哭泣，却体察不到他人的悲苦。他可以听到内心的呼唤，却不知怎样关爱灵魂。

Nowadays, life is moving at such a fast pace; a man's life is not limited to his own village, or his little town. Man moves easily between seas and continents; the number of people he meets is not reckoned in dozens or hundreds, but in tens of thousands.

Under such circumstances, if one needs to communicate and be able to hit it off with strangers at a moment's notice, one must achieve the art of the heart.

Nowadays there is a myriad of books for self-improvement, teaching you how to display your strengths, how to impress at a moment's notice, how to get through those impossible interviews, and once smitten, how to pursue a member of the opposite sex. There are so many skills required to achieve success, so much so that human relationships have become a field of technology, a master craft where certain tricks could work wonders. However, the truth is that to retain good relationships and acquire communication skills, listening is more important than talking.

As far as human development is concerned, we start by learning to listen. I refer to "learning" as a process, because in some cases a person could spend his/her whole life without being able to listen. He may hear the sound of snow falling, but has no feeling for the solemnity of the moment. He could hear the laughter of children, but is not moved by their innocence. He could hear weeping but could not feel for the pain of another human being. He could not help hearing the call of his inner voice, but does not know how to respond to his own conscience.

从婴儿开始，我们就无意识地在听。听亲人的呼唤，听自然界的风雨，听远方的信息，听社会的约定俗成。这是一种模糊的天赋，是可以发扬光大也可以湮灭无闻的本能。有人练出了发达的听力，有人干脆闭目塞听。有很多描绘这种状态的词语，比如“充耳不闻”“置若罔闻”……对“闻”还有歧视性的偏见，比如“百闻不如一见”。

听是需要学习的，它比“说”更重要。如果我们没有听到有关的信息，我们的“说”就是无的放矢。轻率的人，容易下车伊始就哇里哇啦地说，其实沉着安静地听，是人生的大境界。

只有认真地听，你才能对周围有更确切的感知，才能对历史有更深刻的把握，才能把他人的智慧集于己身，才能拓展自己的眼界和胸怀。

读书是一种更广义的倾听。你借助文字，倾听已逝哲人的教诲。你借助翻译，得知远方异族的灵慧。

Starting from infanthood, we unwittingly begin learning to listen. We hear the voices of our parents, the sounds of wind and rain, and news from faraway places. We hear of the mores of the society at large. Hearing is a dubious talent, which could flower out into a real talent, or fade away in obscurity. Some people have nurtured a keen sense of hearing. Others hide themselves in self-imposed deafness.

There are many sayings associated with the concept of hearing, such as “stuff up one’s ears to stop hearing,” “disregard as unreliable hearsay” and so on. The Chinese ideogram for hearing—wen—is sometimes manipulated to carry derogatory overtones, as seen in the expression: “to shut one’s eyes and stop one’s ears,” or in referring to something as “not worth listening to,” or referring to someone or something as: “being heard a hundred times is no match for being seen once.”

Actually, to “listen” is more important than to speak. If we are not aware of certain data, our speaking would be aimless. Frivolous people never stop gabbing. Actually to sit still and listen in silence lifts one to a higher level of consciousness.

Only by listening carefully can one attain a kind of superior awareness, achieve a deeper grasp of history, and learn from other people’s wisdom. Only then can one widen one’s range of thoughts and feelings.

Reading is a form of listening, but on a grander scale. Through the conduit of language, you listen to the wisdom of the past. Through the conduit of translation, you become aware of the wisdom of foreign nations from faraway places.

倾听使人生丰富多彩，你将不再囿于一己的狭隘贝壳，潜入浩瀚的深海。倾听使人谦虚，知道山外有山天外有天。倾听使人安宁，你知道了孤独和苦难并非只莅临你的屋檐。倾听使人警醒，你知道此时此刻有多少大脑飞速运转，有多少巧手翻飞不息。

倾听是美丽的。你因此发现世界是如此五彩缤纷。倾听是幸福的一种表达，因为你从此不再孤单。

倾听是分层次的。某人在特定的时刻，讲了特定的话。只有当我们心静如水，才能听到他的话后之话。年轻人最易犯的毛病是——他明白所有倾听的要素，也懂得做出倾听的姿态，其实呢，他在想着自己待会儿要说的话。他关注的不是述说者，而是自己。“佯听”是很容易露馅的，只要他一开口讲话，神游天外的破绽就败露了。两个面对面述说的人，其实是最危险的敌人。一切都被心灵记录在案。

倾听是老老实实的活儿，来不得半点虚假和做作。倾听是对真诚直截了当的考验。所以，如果你不想倾听，那不是罪过。如果你伪装倾听，就不单是虚伪，而且是愚蠢了。

Listening enriches the lives of people, whisks them out of the shells of their narrow existence and takes them for a dip in the boundless sea of life. Listening is a humbling experience; it makes us aware of mountains beyond mountains and skies beyond skies. Listening brings peace. You learn that loneliness and suffering have landed on others as well as on yourself. Listening makes people alert; you become aware of all the heads that are spinning thoughtfully, and all the hands that are busy at work.

To listen is beautiful. By listening, you discover that the world is so colorful. To listen is to experience happiness, because through listening, you are never lonely.

Listening is a process going through several stages: somebody said something at a certain moment in time, but it is only under certain circumstances when our minds are placid as water do we grasp what had been conveyed beyond the spoken words. As for the young, they know perfectly well what is expected of the listener and they are good at striking the pose of a keen listener, but they are actually preoccupied in sketching out their own contribution to the conversation. It is so easy to see through such pretenses at listening. Because the minute the individual opens his mouth, you can tell that his thoughts had been wandering in a land of their own. Thus, two persons sitting face to face having a so-called conversation are actually each other's worst enemy, leaving a stain in the heart.

To listen is a down-to-earth business, with no room for fancy tricks. To listen is a down-to-earth check on sincerity. If you are not interested in hearing something, it is no big deal. But if you make a pretense of listening, then it is downright stupid.

当我深刻地明白了倾听的本质而不是仅仅把它当成讨好的策略后，倾听就向我展示了它更加美丽的内涵，它无处不在，与我们息息相关。如果你谦虚，以万物为师长，你会听到松涛海啸、雪落冰融，你会听到蚂蚁的微笑和枫叶的叹息。如果你平等待人，你的耐心就有了坚实的基础，你可以从述说者那里获得宝贵的馈赠。这就是温暖的信任和支撑。

年轻的朋友们，让我们学会倾听吧。当你能够沉静地坐下来，目光清澄地注视着对方，抛弃自己的傲慢和虚荣，微微前倾你的身姿，那么你就能听到心与心碰撞的清脆音响，宛若风铃。

I finally realized the nature of listening and stopped seeing it as just a tool for forging relationships. The full nature of the art of listening was finally revealed to me in all its beauty. It permeates our lives in a grand connectedness. If you are humble and accept that everything on earth has something to teach, then you will hear the roaring of oceans and the falling of snow. Your patience will be securely anchored; your listening will be richly rewarded in heartwarming trust and support.

Dear young friends, let us learn the art of listening. When you are able to sit down quietly, inclining slightly forward, your gaze resting on the person in front of you, you yourself totally divested of your *pride and prejudice*, then and only then will you be able to hear the tinkle of two hearts beating together, like bells in the wind.

倾听，是你的魅力

Lend me Your Ear

倾听使人生丰富多彩，你将不再囿于一己的狭隘贝壳，

潜入浩瀚的深海。

等待你的第二颗糖

这是一个寂静的午后。苹果花的香味，弥漫在美国得克萨斯州的一个镇小学的校园里。其中一个班的8个学生，被老师带到了校长室旁边的一间很大的空房里。玻璃窗明晃晃的耀眼，鸟儿飞过的痕迹也能看得清清楚楚，正当学生们强按住内心的好奇，凝神等待着将要发生的一切时，老师领着一个陌生的中年男子走了进来。

他一脸和蔼地来到孩子们中间，给每个人都发了一粒包装十分精美的糖果，并告诉他们：

这糖果属于你的，可以随时吃掉，但如果谁能坚持等我回来以后再吃，那就会得到两粒同样的糖果作为奖励。说完，他和老师一起转身离开了这里。

等待是漫长的，许诺是遥远的，而那颗糖果却真真切切地摆在每个孩子的面前。时间一分一秒地过去了。这颗糖果对孩子们的诱惑也越来越大，伴随着窗外苹果花的芬芳，这种诱惑几乎不可抗拒。

有一个孩子剥掉了精美的糖纸，把糖放进嘴里并发出“啧啧”的声音。受他的影响，有几个孩子忍不住了，纷纷剥开了精美的糖纸。但仍有一半以上的孩子在千方百计地控制着自己，一直等到那个陌生人回来。那是一个比暑假还漫长的40分钟。但陌生人最终实现了自己的承诺，那些付出等待的孩子得到了应有的奖励。

Gratification Deferred

It was a quiet afternoon in a little town in Kansas. Sweet scents of an old apple tree permeated the air of a school yard. There was a class of eight children, and the teacher led them into a large room next to the principal's office. Sunlight shot through the windows, dazzling the eyes of the children. Stains of bird droppings were clearly visible on the glass panes. The children in the room could hardly contain their curiosity, waiting with bated breath for whatever it was that was coming. Finally the teacher led in a middle-aged man, a stranger. Very kindly, the stranger walked up to the children one by one and handed each a beautifully wrapped candy, saying to them: "This piece of candy is yours. If any of you can wait for my return to enjoy it, I will give you two more of the same." Saying so, the man turned and left with the teacher.

It was a long wait, and it was a far-fetched promise. On the other hand, the other candies left behind were staring at them face-to-face. Time ticked by, minute by minute. For the children, their craving for the candies was growing by the minute. Enhanced by the sweet smell of the flowering apple-trees, the temptation was getting beyond control.

One child peeled off the wrappings, put the candy into his mouth and smacked his lips appreciatively. Under this spell, several more children could not hold out any longer, and started unwrapping their candies. But at least half of the children in the room still held out and waited. That forty-minute waiting seemed longer than the summer holidays. Eventually the stranger kept his promise, and the children who had waited got their well-deserved rewards.

事实上，这是一次叫作“延迟满足”的心理实验。后来，那个陌生人跟踪这些孩子整整20年。他发现，能够“延迟满足”的学生，数学、语文的成绩要比那些熬不住的学生平均高出20分。

参加工作后，他们从来不在困难面前低头，总是能走出困境，获得成功。

抵御唾手可得的诱惑，期待通过努力才能得到的许诺，并不是一件容易的事情。对照现实生活，再遇到同样的问题，或许就不难做出正确的选择。

The whole affair had been a psychological experiment called "deferred gratifi-cation." In fact, the stranger followed the children for the next twenty years. He discovered that those children who could stand the test of "deferred satisfaction" did better in maths and languages and beat the other children by 20 grades on average.

Later, when they joined the work force, they never gave up in the face of difficulties, they could always work through the problem and get the job done.

With temptation dangling within easy reach, or a promised reward only to be secured through hard work, it is not easy. But once you have passed the test, it will not be hard later in real life to make the right decision under similar circumstances.

卑微也是我们的朋友

如果你自卑，不要把这视为奇耻大辱。人人都自卑，只是我们战胜自卑的方式不一样。承认自卑是正常的，这就是胜利的第一步。

我常常听到很多人述说自己因为种种理由而自卑，比如个子矮小、家庭贫困、父母双亡或是单亲、受教育的程度太低、不知道某个常识被人耻笑、开运动会买不起新的运动鞋、嗓子太粗不能像夜莺般美妙、头太大了、说话带有明显的乡下口音，等等。

如果说这些在一般人的印象中是弱项，从而成为自卑的理由，那么，我们还比较容易理解。我还听到过有人因为自己太美丽而自卑。那姑娘讲，她付出努力所取得的一切成就，都被人归结为美貌带来的幸运，甚至还有人在话里话外地敲打她是不是运用了某种潜规则。

这个清俊的女生满怀幽怨地说："我为我的相貌而深深自卑。我很想去整容，把自己整得丑陋一些，这样就可以抬起头来做人，人们就会认识到我是一个有内在价值的人。

We are Small Fry, and so What!

If you have a sense of inferiority, don't let it bother you! It is not a disgrace. Actually, everybody has a sense of inferiority. People try to overcome it in different ways. To begin with, let us acknowledge that it is normal to carry within ourselves a sense of inferiority. A good start!

I regularly get tons of letters from people whining about their own sense of inferiority for one reason or another. For instance, they are too short in stature, or short of money; or they are orphaned and live with broken families; or they are under-educated and make themselves a laughing-stock when they betray their ignorance over some commonplace topic, or they cannot afford a new pair of shoes for a sports event; or their voices are too hoarse, obviously unlike the voice of the nightingale; or their heads are too big; or when they open their mouths to speak, they cannot hide their rustic accents, and so on and so forth.

The items listed above could be generally accepted as personal disadvantages and constitute some basis for a sense of inferiority, all of which are understandable.

However, I heard of someone who has an inferiority complex due to her good looks. The girl complained that everything which she had accomplished through her own hard work had been imputed by backbiters to her good looks—some even hinting at shady deals.

With sad resignation, this delicate beauty told me: "I suffer from a deep sense of inferiority due to my looks. I had wanted to get a facelift, to get a plain face, so that I could hold up my head and let people see that I am a person with innate values, not just a pretty face..."

“不骗你，我真的到整形医院去了，可整形师说从来没有接收过这样的病例，他想不出如何操作……”

对于人人都自卑这件事，我是百分百相信。你若是不信，可以抽空看看名人的传记。几乎没有一个名人不谈到自己是自卑的。而且按照咱们上面列举的自卑理由，他们也都是“师出有名”的。

“我不如别人。我自卑，所以，我不停地努力。当年从郑州打到国家队的时候，没有一个人肯定我，他们全说一米五的我打球不会打得如何。为了证明给他们看，我快发了疯，每天都比别人刻苦，我知道我的个子不如别人，别人允许有失败的机会，我没有。我只能赢，所以我打球凶狠，那是逼出来的。后来我成功了，别人又说我没有大脑，只会打球。于是我发疯地学习，英语从不认识字母到熟练地和外国人对话。我不比别人聪明，我还自卑，但一旦设定了目标，就绝不轻言放弃，什么都不用解释，用胜利说明一切！”

这段话是谁说的啊？相信你看完了，就会知道，这是获得过十八个世界冠军，得过四枚奥运金牌的邓亚萍。

也许你要说，这么伟大的人，我们就不好比了。那容我再来录上一段普通人的自卑史。

"To tell you the truth," she continued, "I did go to a beauty salon with my request, but the beautician said he had never heard of such a request, and had no idea how to go about it..."

I absolutely believe that everyone has a sense of inferiority.

If you have doubts, just check the autobiographies of famous people—they invariably suffer from inferiority complex, and can give you a list of their deficiencies.

"I am not as good as the others; I must never stop trying. When I left the Zhengzhou team to join the national team in Beijing, nobody thought much of me. Being only 1.5 meters in height, I would 'never go far' was the general verdict. I knew that I was at a disadvantage in height. To prove myself, I practiced non-stop. Other players could afford to make a mistake and still keep going. Not me. I must keep winning. That is why I play so desperately. I had no second chance.

" I achieved success. Then people started saying that I was a brainless bat-swinger, that playing ball was all I knew. So I took up English, studying like mad. Starting from scratch, I ended up chatting easily with native speakers. I am not smarter than the rest, plus I have a sense of inferiority. But once my goal is set, nothing will stop me. There is no need for words. Victory speaks for itself!"

Now who is this person speaking? By the time you finish reading the passage, you will realize that it is none other than Deng Yaping, the ping-pong athlete who had been world champion for eighteen times, and won four Olympic gold medals.

Now you may complain that it is not fair to make us compare ourselves to such a celebrity. Then let us try the history of an ordinary person and check out her history of self-depreciation.

“我曾经是个非常自卑的人，即使是现在，自卑还常常在，我觉得自己很多地方不如人。我不如A聪明，不如B睿智，不如C有才，不如D貌美如花……我只是一个普通女子，不善言，不会搞各种关系，我只会写字，通过写字证明我自己。感谢我的自卑，它让我越挫越勇，它让我觉得不如别人，让我不敢停步，让我在人生的路上，一路坚强。”

这位女子的文章常常见诸报端，你打开《读者》《青年文摘》等刊物，经常会看到她的文章。

我手边看到的资料说到，因为出演了《色·戒》而再次获奖的梁朝伟，就说自己一直是个非常自卑的人。名人尚且如此，遑论我等俗众!

不要把自卑看得那么可怕，这是人人都享有的一个特点。其实这话说得有语病，既然是人人都有，就不能说是特点了，只能说是常数，对于一个规律性的东西，实在没有害怕的必要，从容对待就是了。因为渺小的人类对于浩瀚的宇宙来说是自卑的。孱弱的婴孩对于伟壮的成年人来说是自卑的。短暂的生命对于无涯的时空来说是自卑的。我们的种种欠缺和无奈，对于光明的期望和理想来说是自卑的。

刚才说了那么多自卑的合理性，并非要大家就对自卑安之若素。其实，你接纳了自卑，你把自卑当成一个朋友，它就会以意想不到的方式来帮助你。

"I used to have an inferiority complex; even now I feel myself to be very inadequate compared to other people. For instance, I am not as smart as A, not as wise as B, not as talented as C, and not as good-looking as D. I am just an ordinary girl, slow of speech and not good at making connections. All I do is putting words together. Through words, I found my own identity. Thanks to my low self-esteem, I never stop trying to improve myself, and never stop working to keep up with the pace of life itself!" Despite her self-depreciation, this young woman's articles often appear in the pages of popular periodicals such as "*The Reader*," or "*Youth Magazine*."

Glancing through the papers, I came across the news that Liang Chaowei, who had won another prize for his role in the new film *Lust and Abstinence*, actually has very low self-esteem. Now if even celebrities have low self-esteem, what can you say for the rest of us small fry?

Well, I suggest that we should not be bothered by a sense of inferiority; it is more or less common to all of us. We should regard it not as an exception, but as a rule. For something that is common to us all, it is futile to fear it or try to get rid of it. Let us just live with it. We are small fry, and so what!

We should accept that our brief lives, seen against a background of boundless time and space, are insignificant, that our imperfections and helplessness, compared to the glorious prospects of the future, are also insignificant.

Having said so much in defense of our sense of insignificance, it does not mean that we small fry should resign ourselves to live with it.

It does not mean that we should live comfortably with our sense of inferiority. Actually once we have accepted it as a friend, it will help us in unexpected ways.

为了战胜自卑，我们就会更加努力。因为自卑的持续存在，我们或许会比较少骄横。因为自卑，我们记得渺小和尊崇，这未尝不是因祸得福呢!

阿尔弗雷德·阿德勒认为，人从一出生就伴随着自卑感，之后需要用一生的时间，去提高自己的技能、优越感和对别人的重要性。

这样看来，卑微也是我们的朋友，卑微里也有不容小觑的力量。

For instance, conscious of our inferiority, we work harder. Likewise, we remember the little things; we remember respect. This is not a bad thing for the small fry.

Alfred Adler had said that self-abasement a follows man the day he is born and he spends a lifetime trying to overcome it and enhance his sense of self-worth.

From this angle, we might say that a sense of inferiority is a friend of us small fry, and that it has a power of its own!

卑微也是我们的朋友

We are Small Fry, and so What!

每个人都是多棱的，即使是一个高尚的人，灵魂中也潜伏着卑微。

你为什么而活着

我有过若干次讲演的经历，在北大和清华，在军营和监狱，在农村土坯搭建的课堂和美国最奢华的私立学校……面对从医学博士到纽约贫民窟的孩子等各色人群，我都会很直率地谈出对问题的想法。在我的记忆中，有一次的经历非常难忘。

那是一所很有名望的大学，约过我好几次了，说学生们期待和我进行讨论。我一直推辞，我从骨子里不喜欢演说。每逢答应一桩这样的公差，就要莫名地紧张好几天。但学校方面很执着，在第N次邀请的时候说，该校的学生思想之活跃甚至超过了北大，会对演讲者提出极为尖锐的问题，常常让人下不了台，有时演讲者简直是灰溜溜地离开学校。

听他们这样一讲，我的好奇心就被激励起来，我说我愿意接受挑战。于是，我们商定了一个日子。

那天，大学的礼堂挤得满满的，当我穿过密密的人群走向讲台的时候，心里涌起怪异的感觉，好像是“文革”期间的批斗会场，不知道今天将有怎样的场面出现。果然，从我一开始讲话，就不断地有条子递上来，不一会儿，就在手边积成了厚厚一堆，好像深秋时节被清洁工扫起的落叶。我一边讲课，一边充满了猜测，不知道树叶中潜伏着怎样的“思想炸弹”。讲演告一段落，进入回答问题阶段，我迫不及待地打开了堆积如山的纸条，一张张阅读。那一瞬，台下变得死寂，偌大的礼堂仿若空无一人。

There is no Meaning to Life....

I do have some experience in public speaking—at Peking University and Tsinghua University, in barracks and prisons, in village classrooms built of beaten earth and in the gleaming halls of prep schools in the US. My audience had ranged from Ph.D. candidates to kids from the slums of New York. I am always frank in saying what I think.

There is one occasion which stands out in my memory. It was a prestigious university and I had turned down the invitation several times running. Public speaking is not my forte. I tend to get nervous several days ahead. But the university kept repeating the invitation, claiming that their students were just as brainy as those of Peking University or Tsinghua University if not more, and that sometimes their challenging questions would send the speaker walking off in a huff. My curiosity was piqued and I decided to accept the challenge.

The hall was packed. As I made my way through the crowd to reach the stadium, my thoughts went back to the "struggle meetings" of the Cultural Revolution, and I wondered what was lying in store for me today. Not surprisingly, the moment I opened my mouth to speak, slips of paper with questions kept arriving, and soon made a pile on the stand. As I talked, my thoughts strayed to the pile before me, and I wondered what startling questions were waiting to confound me. My talk ended, I opened the slips one by one and read the questions aloud, while the packed hall was wrapped in complete silence.

我看完了纸条说：“有一些表扬我的话，我就不念了。除此之外，纸条上提得最多的问题是——人生有什么意义？请你务必说真话，因为我们已经听过太多言不由衷的假话了。”

我念完这个纸条以后，台下响起了掌声。我说：“你们今天提出这个问题很好，我会讲真话。我在西藏阿里的雪山之上，面对着浩瀚的苍穹和壁立的冰川，如同一个茹毛饮血的原始人，反复地思索过这个问题。我相信，一个人在他年轻的时候，是会无数次地叩问自己——我的一生，到底要追索怎样的意义？

“我想了无数个晚上和白天，终于得到了一个答案。今天，在这里，我将非常负责地对大家说，我思索的结果是：人生是没有任何意义的！”

这句话说完，全场出现了短暂的寂静，如同旷野。但是，紧接着就响起了暴风雨般的掌声。

那是我在讲演中获得的最热烈的掌声。在以前，我从来不相信有什么“暴风雨”般的掌声这种话，觉得那只是一个拙劣的比喻。但这一次，我相信了。我赶快用手做了一个“暂停”的手势，但掌声还是绵延了若干时间。

我说：“大家先不要忙着给我鼓掌，我的话还没有说完。我说人生是没有意义的，这不错，但是——我们每一个人要为自己确立一个意义！

Apart from favorable comments on my talk which I put aside, the question which showed up most frequently was, "What is the meaning of life?" One note in particular had pleaded: "Please tell us the truth—we are sick of prevaricating platitudes." As I read that particular note aloud, the audience broke into applause. I said: "This is a good question. I will speak the truth. When I was in Tibet, facing the snowy mountains and the frozen lakes, almost becoming a caveman myself, I had often thought of this question. I believe that when young, one would often ask oneself: 'What is the meaning of the life that I should strive for?' Day and night, I had pondered over the question, and one day I finally found the answer. Today let me tell you in all seriousness my answer to the question: there is no meaning to life..." There was a moment of stunned silence, followed by deafening applause.

It was the most sincere applause that I had ever received. I have always brushed off terms like "deafening applause" as awkward exaggerations. But on that occasion, it was literally true. The applause was thunderous and deafening. I raised my hand for silence, but the applause continued for a while before it quieted down.

I said: "Do not rush to applaud. I have not finished the sentence. I say that life has no meaning. It is up to us—every single one of us—to confer meaning on our own lives!"

“是的，关于人生的意义的讨论，充斥在我们的周围。很多说法，由于熟悉和重复，已让我们从熟视无睹滑到了厌烦。可是，这不是问题的真谛。真谛是，别人强加给你的意义，无论它多么正确，如果它不曾进入你的心理结构，它就永远是身外之物。比如我们从小就被家长灌输过人生意义的答案。在此后漫长的岁月里，谆谆告诫的老师和各种类型的教育，也都不断地向我们批发人生意义的补充版。但是，有多少人把这种外在的框架，当成了自己内在的标杆，并为之下定了奋斗终生的决心？”

那一天结束讲演之后，我听到有同学说，他觉得最大的收获是听到有一个活生生的中年人亲口说，人生是没有意义的，你要为之确立一个意义。

其实，不单是中国的青年人在目标这个问题上飘忽不定，就是在美国的著名学府哈佛大学，也有很多人无法在青年时代就确立自己的目标。我看到一则材料，说某年哈佛的毕业生临出校门的时候，校方对他们做了一个有关人生目标的调查，结果是：百分之二十七的人完全没有目标，百分之六十的人目标模糊，百分之十的人有近期目标，只有百分之三的人有着清晰而长远的目标。

二十五年过去了，那百分之三的人不懈地朝着一个目标坚韧努力，成了社会的精英，而其余的人，成就要相差很多。

我之所以提到这个例子，是想说明，在人生目标的确立上，无论中国还是外国的青年，都遭遇到了相当程度的朦胧或是混沌状态。有人会说，是啊，那又怎么样？我可以一边慢慢成长，一边寻找自己的人生意义啊。我平日也碰到很多青年朋友，诉说他们的种种苦难。我在耐心地听完那些折磨他们的烦心事之后，把他们乞求帮助的目光撇在一旁，我会问：“你的人生目标是什么呢？”

It is true that there is a lot of talk about the meaning of life. Some views are so repetitive that they have become tiresome. But that is not the point. The point is: any "meaning" that others impose on you, no matter how proper, will remain alien unless it has become part of your own psychological make-up. Since childhood, and in later years, through schooling and other forms of education, we have all been brain-fed on the "meaning of life" in various forms and versions. I once came across an item of information about Harvard—leaving campus on Commencement, the graduates were questioned about their plans for the future, and it turned out that 27% of them had no plans, 60% were vague, 10% had short term plans and only 3% had clear-cut plans for the future. Now, twenty-five years later, the 3% have become the cream of the crop, while the rest had done poorly in comparison.

I am using this item of information to point out that nowadays, both in China and worldwide, the young tend to be vague about their prospects. They always claim to be still growing and need time to pin down their own version of the meaning of life.

I often come across young people who are full of complaints about this, that and the other. After hearing through their litany of complaints, I would turn my eyes away from their expectant looks and ask: "What are you living for?"

他们通常会很吃惊，好像怀疑我是否听懂了他们的愁苦，甚至恼怒我为什么对具体的问题视而不见，而盘问他们如此不着边际的空话。更有甚者，以为我根本就没有心思听他们说话，自己胡乱找了个话题来搪塞。

我会迎着他们疑虑的目光，说："请回答我的这个问题，你为什么而活着呢？"

年轻人一般会很懊恼地说："这个问题太大了，和我现在遇到的事没有一点关联。"我会说："你错了。世上的万事万物都有关联。有人常常以为心理上的事只和单一的外界刺激有关，就事论事，其实心理和人生的大目标有着纲举目张的紧密接触。很多心理问题，实际上都是人生的大目标出现了混乱和偏移。"

举个例子。一个小伙子找到我，说他为自己说话很快而苦恼，他交了一个女朋友，感情很好。但女孩子不喜欢他说话太快。一听他口若悬河滔滔不绝地说个没完，女孩就说自己快变成大头娃娃了，还说如果他不改掉这毛病，就不能把他引荐给自己的妈妈，因为老人家最烦的就是说话爱吐唾沫星子的人。

"你说我怎么才能改掉说话太快的毛病？"他殷切地看着我，闹得我都觉得如果不帮他这个忙，简直就成了毁掉他一生爱情和事业的凶手。

我说："你为什么要讲话那么快呢？"

他说："如果慢了，我怕人家没有耐心听完我的话。您知道，现在的社会节奏那么快，你讲慢了，人家就跑了。"

我说："如果按照你的这个观点发挥下去，社会节奏越来越快，你岂不是就得说绕口令了？你的准丈母娘就不是这样的人啊，她就喜欢说话速度慢一点并且注意礼仪的人啊。"

They would act shocked, suspecting that I had not heard through their lists of complaints and was badgering them with highfalutin rhetoric. Others go even further, suspecting that I had no interest in their problems, so they wander off to another subject. I would look them in the eye and ask: "Please answer me. What are you living for?" Most young people would be annoyed and say: "The question covers too wide a field and has nothing to do with my immediate problem." I would say "You are wrong. Everything in the world is interconnected. People assume that their psychological problem has to do with a single shock from the outside world, and they deal with it as such. Actually one's psychological make-up has much to do with one's overall goal in life. Many psychological problems stem from confusion and wrong turns regarding one's goal in life."

I gave one example. A young fellow looked me up with a problem—he talked too fast. He found a girlfriend and they got along fine, but the girl was bothered by his speed talk. His words, gushing forth as in a waterfall, drummed on her head until it was about to burst. She said she could not take him home to meet her parents—her mother does not tolerate people who shoot off specks of spit as they speed talk!

"How can I get rid of this troublesome habit?" he looked at me appealingly, as if I would be responsible for wrecking his marriage and his career if I did not help.

I asked him: "Why do you need to talk so fast?"

He said: "If I slow down, I will lose people's attention. You know how it is nowadays—the pace of life is so fast!"

"According to your kind of thinking, the society will be spinning faster and faster until we will be forced to communicate in laconic cross-talk! Actually, even by your own account, your prospective mother-in-law is not that intimidating. I presume she prefers speech to be more deliberate, with some regard to manners as well."

他说："好吧，就算你说的这两种人都可以并存，但我还是觉得说话快一些，比较占便宜，可以在单位时间内传达更多的信息。"

我说："那你的关键就是期待别人能准确地接受你的信息。你以为只有快速发射信息才是唯一的途径。你对自己的观点并不自信。"

他说："正是这样。我生怕别人不听我的，我就快快地说，多多地说。"

当他这样说完之后，连自己也笑起来。我说："其实别人能否接受我们的观点，语速并不是最重要的。而且，你能告诉我，你为什么这样在意别人是否能接受你的观点？"

这个说话很快的男孩突然语塞起来，忸怩着说："我把理想告诉你，你可不要笑话我。"

我连连保证绝不泄密。他说："我的理想是当一个政治家。所有的政治家都很雄辩，你说对吧？"

我说："这咱们就接触到了问题的实质。要当一个政治家，第一要自信。他们的雄辩不是来自速度，而是来自信念。一个自信的人，不论说话快还是慢，他们对自我信念的坚守流露出来，会感染他人。我知道你有如此远大的理想，这很好。你要做的事，不是把话越说越快，而是积攒自己的力量，让自己的信念更加坚强。"

那一天的谈话到此为止。后来，这个男生告诉我，他讲话的速度已慢了下来，也被批准见到了自己的准丈母娘，听说很受欢迎。

"Okay," he conceded, "Let's say that both ways of speaking are equally acceptable, but I still prefer to be faster—at least you have the advantage of conveying more info within a fixed time limit."

"In which case," I retorted, "you take it for granted that people understand and accept your views. And to achieve this end, you feel you must send out all your messages as quickly as possible. It denotes a lack of self-confidence."

"That is true," he conceded, "I worry that people do not get the message, so I do the 'fast-talk' to pack in as much info as possible."

As he said this, he laughed at himself.

"Actually," I pointed out, "whether people accept or reject our views do not depend on the speed of our speech. Besides why should you bother whether people accept your views or not?"

The young fellow became embarrassed, and said: "I will tell you the reason if you promise not to laugh at me." Of course I promised, and he said: "I want to be a politician, and all politicians are orators, aren't they?"

"Now we have arrived at the crux of the problem—you want to go into politics. For that, the first thing you need is self-confidence. Politician's oratory skills do not come from speed, it comes from self-confidence. A self-confident person—never mind his speaking skills—will inspire his audience by the strength of his convictions, and his speeches will naturally be rousing. It is good that you have an ideal to cling to. But what you need is not speed-talk. You should strengthen your convictions and build up your base."

And thus we ended the conversation. Later, this young man told me that his speech had slowed down, and that he met his prospective mother-in-law and had actually received a warm welcome!

这边刚刚解决了一个说话快的问题，紧接着又来了一位女硕士，说自己的心理问题是讲话太慢，周围的人都认为她有很深的城府，不敢和她交朋友，以为在她那些缓慢吐出的话语背后，隐藏着怎样的阴谋。

“我试了很多方法，却无法让自己说话快起来，烦死了。”她慢吞吞地对我这样说，语速的确有一种压抑人的迟缓，好像在话的背后还隐藏着另一句话。

我看她急迫的神情，知道她非常焦虑。

我说：“你讲每一句话是否都要经过慎重的考虑？”

她说：“是啊。如果不考虑，讲错了话，谁负得了这个责？”

我说：“你为什么特别怕讲错话？”

女硕士说：“因为我输不起。我家庭背景不好，家里有人犯了罪，周围的人都看不起我们；家里很穷，从小靠亲戚的施舍我才能坚持学业。我生怕一句话说差了，人家不高兴，就不给我学费了。所以，连问一句‘你吃了吗？’这样中国最普通的话，我也要三思而后行。我怕人家说，你连自己的饭都吃不饱，也配来问别人吃饭问题。”

听到这里，我说：“我明白了。你觉得自己的每一句话都可能引致他人的误解，给自己造成不良影响。”

女硕士连连说：“对对，就是这样的。”

我笑了，说：“你这一句话说得并不慢啊。”

她说：“那我是相信你不会误会我。”

Just as I had dealt with the fast-talking young man, a PhD candidate arrived, who said that her psychological problem was that her speech was too slow. People keep a distance from her, she said, suspecting that she is too 'deep,' judging by the deliberate way she picked her words.

"I had tried, but I just cannot change the way I speak. It is such a nuisance," she confessed. Her speech did indeed carry with it a sense of depression, as if one word was being stifled beneath every new word that she was slowly releasing.

I could see that the young woman was truly depressed, and I felt the urgency of her problem.

I asked her: "Do you deliberate over every word you utter?"

"Of course," she said, "If you pick a wrong word without thinking, who is going to be responsible, except you yourself?"

"But why are you so anxious about the choice of words?"

"I can't afford to make a mistake. My family background is bad. A member of my family has broken the law, so our whole family is despised in the neighborhood. Besides, we are not well off. Since childhood, my education had been supported by relatives. I am afraid of speaking out of turn, offending the relatives and losing their support. I am even too shy to use the usual Chinese greeting 'Have you eaten?' People might say: 'You can't even feed yourself; what's the point of asking if we have eaten?'"

"I understand," I said, "You worry that every word you say is liable to be twisted and used against you?"

She agreed immediately: "That's exactly it!"

"Now that was a quick answer!" and I laughed.

"Well, I am confident that you will not misunderstand."

我说：“这就对了。你说话速度慢，不是一个技术性的问题，是你不能相信别人。你是否准备一辈子都不相信任何人？如果是这样，我断定你的讲话速度是不会改变的。如果你从此相信他人，讲话的速度自然会比较适宜，既不会太慢，也不会太快，而是能收放自如。”

那个女生后来果然有了很大的改变，她的人际关系也有了进步。

今天我们从一个很大的目标谈起，结果要在一个很小的地方结束。我想说，一个人的心理是一座斗拱飞檐的宫殿，这座宫殿的基础就是我们对自己人生目标的规划和对世界对他人的基本看法。一些看起来是技术和表面的问题，其实内里都和我们的基本人生观有着千丝万缕的联系。心理问题切不可头痛医头脚痛医脚，那样如同创可贴，只能暂时封住小伤口，却无法从根本上让我们的精神强健起来。

“That’s right. Your slow speech is not a physical problem. It is because you do not trust other people. Are you going to spend your whole life distrusting everybody around you? If yes, I can assure you that your speed in speaking will never pick up. If you begin to trust people your speed in speaking will be normal, neither too fast nor too slow.”

The young girl later did improve her speech, as well as her relationship with her peers.

I rounded up my talk by saying:

“Today we have started out on a big topic, and will come to a stop in a very small place. I want to say that a person’s psychology is like a castle with flying eaves. The base for this castle is your plan for life, as well as your view of other people and of the world in general. Some problems which seem at first glance to be technical are in fact linked to our world view. Psychological problems cannot be dealt with by keeping to the surface—that will be like sticking plaster, used for minor aches and pains. It cannot help in the fundamentals, which is to infuse strength into our own spirit to make us strong.”

轰毁你心中的魔床

魔鬼有张床。它守候在路边，把每一个过路的人，揪到他的魔床上。魔床的尺寸是现成的，路人的身体比魔床长，它就把那人的头或是脚锯下来。那人的个子矮小，魔鬼就把路人的脖子和肚子像拉面一样抻长……只有极少的人天生符合魔床的尺寸，不长不短地躺在魔床上，其余的人总要被魔鬼折磨，身心俱残。

一个女生向我诉说："我被甩了，心中苦痛万分。他是我的学长，曾每天都捧着我的脸说，你是天下最可爱的女孩。可说不爱就不爱了，做得那么绝，一去不回头。我是很理性的女孩，当他说我是天下最可爱的女孩的时候，我知道我姿色平平，担不起这份美誉，但我知道那是出自他的真心。那些话像火，我的耳朵还在风中发烫，人却大变了。我久久追在他后面，不是要赖着他，只是希望他拿出响当当、硬梆梆的说法，给我一个交代，也给他自己一个交代。

"由于这个变故，我不再相信自己，也不相信他人。我怀疑我的智商，一定是自己的判断力出了问题。如此至亲至密，说翻脸就翻脸，让我还能信谁？"

The Demon Bed

The Demon has a bed. It would bring the bed to the edge of the highway and snatch up people passing by. It would force those passers-by down on the bed. If the poor creatures could not fit into the bed, the Demon would cut off their heads or saw off their legs to create a perfect fit. If a passer-by was short, the Demon would stretch its neck and pull at its tummy until it was the right length to fill the bed. Very few people happen to be the right size to make a perfect fit. Practically no one passing by could escape the physical and mental torture of the Demon's Procrustean bed.

A girl told me: "He jilted me. It hurts like hell. We were schoolmates, he is a grade above me. Every day, he would cup my chin in his palm and say 'You are the loveliest girl in the world.' And then he dropped me with a finality that knows no turning back. I am rational. When he said that I was the loveliest girl in the world, I myself knew that it was an exaggeration, that my looks do not merit that kind of praise. But I knew that at the time, he meant what he said. My ears are still tingling, but now he has become another man. I nagged him for quite a while. Not that I won't give him up. The fact is, I want a clear-cut explanation for his change of heart, an explanation resounding with conviction and eloquently delivered—for his own sake and for mine.

"Having gone through this ordeal, I lost confidence in myself and I lost trust in other people. I am beginning to doubt my own IQ—I must have been lacking in judgment, else how can I explain his sudden change of heart when we had been so close? How can I ever trust anybody again?"

女生叫萧凉，萧凉说到这里，眼泪让围巾的颜色一片片变深。失恋的故事，我已听过成百上千，每一次，不敢丝毫等闲视之。我知道有殷红的血从她心中坠落。

我对萧凉说："这问题对你，已不单单是失恋，而是最基本的信念被动摇了，所以你沮丧、孤独、自卑还有愤怒的莫名其妙……"

萧凉说："对啊，他欠我太多的理由。"

我说："人是追求理由的动物。其实，所有的理由都来自我们心底的魔床——那就是我们对一些问题的看法和观念。它潜移默化地时刻评价着我们的言行和世界万物。相符了，就皆大欢喜，以为正确合理。不相符，就郁郁寡欢怨天尤人。"

这种魔床，有一个最通俗最简单的名字，就叫作"应该"。有的人心里摆得少些，有三个五个"应该"。有的人心里摆得多些，几十个上百个也说不准，如果能透视到他的内心，也许拥挤得像个卖床垫的家具城。

魔床上都刻着怎样的字呢?

萧凉的魔床上就写着"人应该是可爱的"。我知道很多女生特别喜欢这个"应该"。热恋中的情人，更是三句话不离"可爱"。这张魔床导致的直接后果，就是我们以为自己的存在价值，决定于他人的评价。如果别人觉得我们是可爱的，我们就欢欣鼓舞，如果什么人不爱我们了，就天地变色日月无光。

The girl was called Xiao Liang. As she spoke, tears dripped down her cheeks, leaving stains on the scarf she was wearing. A story of love betrayed, of which I have heard hundreds of times. It is not something to be treated lightly. I knew that the girl's heart was bleeding.

I said: "To you, this is not just a problem of disappointed love. The basic convictions that you have always held on to are now shaken. You are left with humiliation, loneliness, self-defeat, anger and incomprehension."

Xiao Liang agreed: "That's exactly it! He owes me a lot of explaining!"

I said: "The human being belongs to a species that looks for reasons. Actually, all reasons come from a Demon bed hidden in the bottom of our hearts! I am referring to our view of things and the personal convictions that we hold on to. Silently, they pass judgments on our behavior, on the world, and everything in it. If reality fits into our convictions, then all is well and good. If reality refuses to fit into our convictions, then we are unhappy and start playing the blame game." This is what is meant by Demon bed, and this kind of bed has a very simple name—*should*.

People's hearts are littered with the word *should*: a handful in some cases, more in other cases, and for some people literally in the hundreds—their hearts are taken over by Demon beds under the name *should*. What are the words on those Demon beds? Well, on Xiao Liang's Demon bed are the words "People should be lovable." I know many girl students who love the word *should*. As for ardent lovers, they always have the word *lovable* hanging on their lips.

The direct consequence of this kind of Demon bed in our hearts is that our own sense of self-worth rests on other people's approval. If other people think us "lovable," we are elated. If someone falls out of love with us, then the sun and the moon are plunged into darkness.

很多失恋的青年，在这个问题上百思不得其解，苦苦搜索“给个理由”。如果没有理由，你不能不爱我。如果你说的理由不能说服我，那么就只有一个理由，就是我已不再可爱，一定是我有了什么过错……很多失恋的男女青年，不是被失恋本身，而是被他们自己心底的魔床，锯得七零八落。残缺的自尊心在魔床上火烧火燎，好像街头的羊肉串。

要说这张魔床的生产日期，实在是年代久远，也许生命有多少年，它就相伴了多少年。最初着手制造这张魔床的人，也许正是我们的父母。当我们还是婴儿的时候，那样弱小，只能全然依赖亲人的抚育。如果父母不喜欢我们，不照料我们，在我们小小的心里，无法思索这复杂的变化，最简单的方式，我们就以为是自己的过错。必是我们不够可爱，才惹来了嫌弃和疏远。特别是大人们的口头禅“你怎么这么不乖？如果你再这样，我就不喜欢你了……”凡此种种，都会在我们幼小的心底，留下深深的印记。那张可怕的魔床蓝图，就这样一笔笔地勾画出来了。

有人会说，啊，原来这“应该如何如何”的责任不在我，而在我的父母。其实，床是谁造的，这个问题固然重要，但还不是最重要的。心理学家弗洛伊德说过，一个孩子，即使在最慈爱的父母那里长大，他的内心也会留有很多创伤。（大意如此。原谅我一时没找到原文，但意思绝对不错。）我们长大之后，要搜索自己的内心，看看它藏有多少张这样的魔床，然后亲手将它轰毁。

Many young people who are jilted cannot get over their predicament; they demand a reason. According to them, if the other party cannot give a reason, then he or she has no right to stop being in love. If the reason offered is not convincing, then there is only one explanation: "I have stopped being lovable" or "I must have done something wrong."

Thus, many young people disappointed in love are actually struck down not by the disappointment itself but by the Demon bed hidden in his/her heart. Looking in vain for a reason, they are cut and hacked at by the Demon bed called *should*.

Coming back to the Demon bed—perhaps it is ancient, perhaps as old as Life; actually it is a part of Life. Perhaps it was our parents who had created this Demon bed with their own hands. At the time, we were still babies, little weaklings, totally relying on our parents to raise us. If our parents stopped loving us and taking care of us, our little hearts would not be able to understand the reason for the change. The easiest way out is to blame ourselves, to accept that we are not lovable—thus leading to alienation from our parents. The alienation is nailed down by words hanging on the lips of our parents: "Why don't you behave? If you continue like this, we will not love you anymore." All of this will leave a deep impression on our little hearts. It is thus that the blueprint of the terrible Demon bed gradually takes shape.

People will say: "I am not responsible for the pressure that had shaped me into 'so and so.' It does not lie with me, but with my parents."

Although it is important to nail down the answer to the question as to who had created the Demon bed, actually it is not of paramount importance. The psychologist Freud had said that even if a child had grown up with loving parents, his or her little heart still carried wounds. When we grow up, we should search our own hearts for the Demon bed hidden therein, and destroy it with our own hands.

一位男青年说：“我很用功，我的成绩很好。可是我不善辞令，人多的场合，一说话就脸红。我用了很大的力量克服，奋勇竞选学生会的部长，结果惨遭败北。前景黑暗，这可不是个好兆头，看来我一生都会是失败者。”于是，他变得落落寡合，自贬自怜，头发很长了也不梳理，邋遢着独往独来的，好似一个旧时的落魄文人。大家觉得他很怪，更少有人搭理他了。

他内心的魔床就是：我应该是全能的。我不单要学习好，而且样样都要好。我每次都应该成功，否则就一蹶不振。挫折被放在这张魔床上翻身反复比量，自己把自己裁剪得七零八落。一次的失败就成了永远的颓势，局部的不完美就泛滥成了整体的否定。

一个不美丽的大学女生每天顾影自怜。上课不敢坐在阶梯教室的前排，心想老师一定只愿看到“养眼”的女孩。有个男生向她表示好感，她想我不美丽，他一定不是真心。如果我投入感情，肯定会被他欺骗，当作话柄流传。于是，她斩钉截铁地拒绝了他，以为这是决断和明智。找工作的时候，她的简历写得很好，屡屡被约见面试，但每一次都铩羽而归。她以为是自己的服饰不够新潮、化妆不够到位，省吃俭用买了高级白领套装外带昂贵的化妆品，可惜还是屡遭淘汰……她耷拉着脸，嘴边已经出现了在饱经沧桑的失意女子脸上才可看到的像小括弧般的竖形皱纹。

A young man had once told me, "I am studious; my grades are good, but I am not good at relationships. I blush and am tongue-tied in company. To overcome this shyness, I competed for the position of President of the Students' Union, but I lost." He said: "It is a bad sign. My future is dark. I am doomed to be a loser for life." Thus the young man became morose, and gave up trying. His hair was uncut; he dragged himself here and there like a member of the literati from the decadent old days. People thought him strange and soon stopped talking to him. The Demon bed consciousness in him murmured: "I should be an all-rounded person—not only a good student, but outstanding in every way. I should always win, or else I would become a dismal failure." Setbacks appeared on his Demon bed, and after turning this way and that, he was convinced that he had indeed become a dismal failure. Thus one failure laid the grounds for a lifetime of depression. One partial imperfection led to his total denial of his own worth. Thus worked the Demon bed in him, the thing called *should*.

A college girl who was not beautiful looked at her own shadow and pitied herself. In class she dared not sit on the first row, thinking that professors prefer to feast their eyes on pretty girls. When a boy showed interest in her, she reminded herself that she was not beautiful and she questioned his sincerity. She feared that if she accepted his advances and later got jilted, she would be a laughing stock. So she refused his advances, thinking that she did the right thing. When looking for a job, her CV was impressive, and she got many interviews, but was always rejected. She thought it was because she was not fashionably dressed, so she saved on food to pay for nice clothes and expensive cosmetics. But she was still rejected. She goes around with that hangdog look and little wrinkles lined up on her face so typical of losers.

如果允许我们走进她枯燥的内心，我想那里一定摆着一张逼仄的小床。床上写着："女孩应该倾国倾城，应该有白皙的皮肤，应该有挺秀的身躯，应该有玲珑的曲线，应该有精妙绝伦的五官……如果没有，她就注定得不到幸福，所有的努力都会白搭，就算碰巧有一个好的开头，也不会有好的结尾。如果有男生追求长相不漂亮的女孩，一定是一个陷阱，背后必有狼子野心，切切不可上当……"

很容易推算，当一个人内心有了这样的暗示，她的面容是愁苦和畏惧的，她的举止是局促和紧张的，她的声音是怯懦和飘忽的……她在情感和事业上成功的概率极低，到了手的幸福不敢接纳，尚未到手的机遇不敢追求，她的整个形象都散射着这样的信息——我不美丽，所以，我不配有好运气！

讲完了暗淡的故事，擦拭了委屈的泪水，我希望她能找到那张魔床，用通红的火将它焚毁。

谁说不美丽的女子就没有幸福？谁说不美丽的女子就没有事业？谁说命运是个好色的登徒子？谁说天下的男子都是以貌取人的低能儿？

心中的魔床有大有小，有的甚至金光闪闪，颇有迷惑人的能量。我见过一家证券公司的老总，真是事业有成、高大英俊，名牌大学洋文凭，还有志同道合的妻子，活泼聪颖的孩子……一句话，简直人所有的他都有，可他寝食无安，内心的忧郁焦虑非凡人所能想象，不知是什么灼烤着他的内心。

If we look into her heart, we are sure to find that Demon bed written over with the words: "Girls *should* be beautiful, with everybody lying at her feet in adoration. She *should* have tender skin, a slim body with curves in the right places, and beautiful feathers for decoration. If lacking in any of the above, she will be doomed to unhappiness, and all her attempts at a relationship will be wasted. Even if she starts well, she will never amount to much... "

"If a boy pursues a girl who is not good-looking," the Demon bed's voice commented, "It is certain to be a trap. There must be something sinister behind it. One *should* take care not to fall into a trap!"

One can easily imagine, with such a psychological burden, a girl will certainly look scared and pitiful. Her movements will be nervous and jerky; her voice will be weak and low. She will keep her eyes down and her gaze will be shifty. She has little chance of succeeding either at work or in her emotional life. She will be too timid either to snatch at happiness within reach or to pursue happiness on her own initiative. Her whole figure sends out signals that she is not beautiful and does not deserve happiness. After wiping her tears and finishing her sad story, I hope she will locate her Demon bed called *should*, and throw it into a burning fire.

Who said that plain-looking girls have no right to happiness? Who said that plain-looking girls have no chance at a career? Who said that Fortune is a rake, in pursuit of beauty? Who said that all men are morons, pursuing a pretty face?

As for the Demon beds in each of us, they are of different sizes and descriptions, some very dazzling in rays of gold. I once came across the big boss of a company, who seemed to have everything a man could wish for: a successful career, an imposing figure, a degree from a famous university, a wife who was a perfect partner, and perfect children, smart and quick-witted—in a word, he was blessed with everything that a man could wish for, but he was plagued by unease, carrying a burden of anxieties beyond endurance. No one knew what was gnawing at his heart.

“我总觉得这一切不长久。人无远虑，必有近忧。水至清则无鱼，谦受益满招损。我今天赚钱，日后可能赔钱。妻子可能背叛，孩子可能车祸。我也许会突患暴病，世界可能会地震火灾飓风，即使风调雨顺，也必将会有人祸，比如‘911’……我无法安心，恐惧追赶着我的脚后跟，惶恐将我包围。”他眉头紧皱地说。

我说：“你极度的不安全。你总在未雨绸缪，你总在防微杜渐。你觉得周围潜伏着很多危险，它们如同空气看不着、摸不到但却无所不在、无所不能。”

他说：“是啊。你说得没错。”

我说：“在你内心，可有一张魔床？”

他说：“什么魔床？我内心只有深不可测的恐惧。”

我说：“那张魔床上写着，人不应该有幸福。只应该有灾难。幸福是不真实的，只有灾难才是永恒。人不应该只生活在今天，明天和将来才是最重要的。”

他连连说：“正是这样。今天的一切都不足信，唯有对将来的忧患才是真实的。”

我说：“每个人都有过去和将来。对我们来讲，无论过去发生过什么，都已逝去。无论你对将来有多少设想，都还没有发生。我们活在当下。”

由于幼年的遭遇，他是个缺乏安全感的人。惊惧射杀了他对于幸福的感知和欣赏。只有销毁了那张魔床，他才能晒到金色的夕阳，听到妻儿的欢歌笑语，才能从容镇定地面对风云，即使风雨真的袭来，也依然轻裘缓带玉树临风。

"I don't think all this will last. If one does not look far out, disaster will strike home. Water, if too clear, will not support fish life. Scarcity promotes charity, and fullness is bound to attract misfortune. I am making money today, but I may lose it tomorrow. My wife may betray me and my son may have a road accident. I may be struck down by a sudden illness. There may be an earthquake, or a fire or a whirlwind sweeping everything away. Even if the weather is benign, there may still be man made disasters such as the 911 Terrorist Attack. Anyway, I cannot find peace. Fear is always at my heels, entangling me!" he kept saying, his brows knit in anxiety.

I told him that he was lacking in a sense of security, frightened at unknown evils, and always looking out for what might befall him. Indeed; he always felt himself living in the midst of imminent disaster unknown and unseen, ready to pounce on him at any moment.

He said, yes, you are right.

I asked, "Is there a Demon bed inside your consciousness?"

"What Demon bed! All I have is fathomless fear!"

I told him: "On that Demon bed are those words: 'Men do not deserve happiness. Men only deserve disaster. Happiness is unreal, disaster is everlasting. Men should not just live in the present. Tomorrow, and tomorrow, is of first importance.'"

He hastened to agree, saying: "Exactly. Nothing lying before our eyes today is worthwhile. The only thing genuine is our anxiety for the future."

Because of what had happened to him in his youth, the man was totally without sense of security. Fear had destroyed his enjoyment of happiness. Only by destroying his Demon bed can he enjoy a golden sunset, hear the happy sounds of laughter of his wife and children, and face squarely what the future has in store. Even if disaster strikes, he will be standing upright firm and strong to brave the storm.

说穿了，魔床并不可怕，当它不由分说就宰割着你的意志和行为之时，面对残缺，我们只有悲楚绝望。但当我们撕去了魔床上的铭文，打碎了那陈腐的“应该”，魔力就在一瞬间消失。随着魔床轰塌，代之以我们清新明朗的心态。

魔由心生。时时检点自己的心灵宝库，可以储藏勇气，可以储藏智慧，可以储藏经验和教训，可以储藏期望和安慰，只是不要储藏“应该”。

Coming down to brass tacks, it is not the Demon bed in us that is to be feared. Of course, if it overwhelms you and controls your will power, you are plunged into despair. But when we shake off the Demon bed and smash the varieties of *should* that we used to defer to, the magic melts away. With the breakdown of the Demon bed in us, our natural frame of mind will be restored.

Demons reside in the heart. We should often check into the deposit of our hearts' treasures. Right now are stored courage, and experience, and lessons at a price and hope and consolation. But let us never store *should* and all that it stands for. Never let it take over our minds!

轰毁你心中的魔床

The Demon Bed

有一颗大心，才盛得下喜怒，输得出力量。

生活就是泥沙俱下，

就是鲜花和荆棘并存，

尊重生活本来的面目，

接受一个完整统一的真实世界。